Managing Major Hazards

Managing Major Hazards

The lessons of the Moura mine disaster

Andrew Hopkins

Routledge
Taylor & Francis Group
LONDON AND NEW YORK

First published 1999 by Allen & Unwin

Published 2020 by Routledge
2 Park Square, Milton Park, Abingdon, Oxon OX14 4RN
605 Third Avenue, New York, NY 10017

Routledge is an imprint of the Taylor & Francis Group, an informa business

National Library of Australia
Cataloguing-in-Publication entry:

Hopkins, Andrew, 1945– .
Managing major hazards: the lessons of the Moura mine disaster.

Bibliography.
Includes index.

1. Coal mine accidents—Queensland—Moura Region. 2. Mine safety—Queensland—Moura Region. 3. Disasters—Case studies. 4. Emergency management. I. Title. (Series: Studies in society).

363.11962233409435

Set in 10/11.5pt Palatino by DOCUPRO, Canberra

ISBN-13: 9781865087023 (pbk)

CONTENTS

For all workers whose jobs put their lives at risk

ACKNOWLEDGMENTS

This book is published with the assistance of the Construction, Forestry, Mining and Energy Union, the National Occupational Health and Safety Commission, South Australian Workcover, NSW Workcover, the Queensland office of Workplace Health and Safety and the Australian Occupational Health and Safety Trust.

I am grateful to Chris Ellicott and Frank Roxborough, for detailed comments on an earlier draft. They of course bear no responsibility for the final version.

Pronunciation note: Moura rhymes with Now-ra

ABBREVIATIONS

ARCO	Atlantic Ridgefield Company
BASI	Bureau of Air Safety Investigation
BHP	Broken Hill Proprietary
BHPAC	BHP Australia Coal
CO	carbon monoxide
DMR	Department of Mineral Resources
DPIE	Department of Primary Industries and Energy
EFTPOS	Electronic Funds Transfer at Point of Sale
GDP	Gross Domestic Product
GFT	general failure types
HRO	high-reliability organisation
HRT	high-reliability theory
HSE	Health and Safety Executive
IC	Industry Commission
LTIFR	lost time injury frequency rate
NASA	National Aeronautical and Space Administration
NOHSC	National OHS Commission
NOSA	National Occupational Safety Association
NSWDMR	New South Wales DMR
OHS	occupational health and safety
QA	quality assurance
RSI	repetition strain injury
SA	Standards Australia

Glossary

deputy The lowest official in the management hierarchy. Deputies are in immediate charge of work crews at the coal face. They are responsible for face operations and for safety.

development phase (of a panel) The construction of a rectangular network of tunnels throughout the length and breadth of the panel.

explosive range When the concentration of methane in a mine's atmosphere is between 5 and 15 per cent it is explosive, that is, it could be ignited by a spark (assuming an otherwise normal atmosphere). This is called the explosive range.

extraction phase (of a panel) After the development phase is completed, miners retreat to their starting point, extracting as much coal as possible from the blocks left standing in the development phase. They must leave sufficient coal standing to protect themselves from roof fall as they retreat.

goaf A mined-out area of a panel, subject to roof fall and normally a no-go area for miners.

heating A condition under which the low-temperature oxidation of coal is such that the heat generated is not sufficiently dissipated by cooling factors, such as the mine ventilation, to prevent the temperature of a coal mass increasing. The increase of temperature in turn results in an increase in the rate of oxidation, which results in an increase in temperature, and so on . . . The term heating may cover a range of conditions: from the start of a coal mass becoming warmer; to a condition when that same mass could provide a source of ignition (Windridge 1996:ADi).

panel A section of the mine that is being actively mined (or has been mined out). As each panel is completed all ventilation to the panel ceases and it is sealed off. (The word 'panel' has a second meaning in this book: the group of people appointed to conduct the inquiry.)

registered mine manager A person appointed pursuant to section 50 of the *Queensland Coal Mine Act* of 1925 by the owner of the mine, such manager being responsible for the control, management and direction of the mine.

sealing The construction of walls at certain tunnel intersections to isolate the panel from the rest of the mine.

spontaneous combustion see heating.

1

INTRODUCTION

Things are always going wrong in organisations. Information fails to get to people who need it; erroneous or counterproductive beliefs undermine good decision making; things fail to happen because no one actually has responsibility to make them happen; and people follow their own agendas at the expense of the organisation's goals. This is all routine. And normally the consequences are not too serious. The organisation might not be as efficient as possible, and the product or service might be less than optimal. But errors, if they are serious enough, will be corrected and no great harm is done. Thus, organisations lumber on, and the ways in which they are failing remain unexposed.

But when a disaster occurs, when people are killed, the organisation concerned is placed under a spotlight. External authorities conduct detailed inquiries which may generate thousands of pages of transcript evidence. In the process, the workings of the organisation are laid bare and its failings, normally hidden from view, are open to scrutiny. Disasters therefore offer an unparalleled opportunity to study the workings of an organisation and to identify where things are going wrong.

This book examines a particular disaster in order to shed light on the way things go wrong in organisations. It is a study of an explosion in a BHP coal mine at Moura in central Queensland in 1994 in which eleven men lost their lives. The official inquiry provided a fascinating insight into the organisational failures that led to the disaster, failures that we can assume are present in many other contexts and that will continue to give rise to disasters unless ways can be found to overcome them. Here then is the practical motivation of this book: it is an attempt to identify the lessons of the Moura disaster. Because this is a study of organisational processes, these lessons go beyond the particular case of underground coal mining; they are relevant to the control of catastrophic risks in a wide variety of contexts.

Controlling the risk of explosion in underground coal mining in Australia is an urgent task. In a 22-year period, from 1972 to 1994, there were no fewer than five methane gas explosions causing major loss of life. A total of 67 men died in these explosions. The impact on the town of Moura, population 3000, has been particularly devastating. Three explosions have occurred in mines in the immediate vicinity of the town over a 19-year period, killing a total of 36 miners. Here are the words of one Moura resident interviewed some time after the third explosion (Background Briefing, 25/4/95).

> This morning I did the same thing I did then. You put your shoes on, you go for a walk, you cry, you go past the Kianga Hall, the monument to the men from the first one, and you look at the statue—it's a monument to all the men in the second one—and it's happened again. And then it's the anger. You get so angry . . .

Anger was not the initial response to the 1994 disaster. The first miner at the pit top to discover what had happened appeared in a doorway to tell his mates. One noted that 'all the colour was drained out of his face and he was actually grey, his face was grey'. What he said was: 'We've lost her, mate, we've bloody lost her.' There is no anger in these words, only a certain fatalism. It was as if he was saying that mines are dangerous and perhaps even temperamental places, beyond the capacity of mere men to control.

Or again, consider this response. Shortly after the explosion, a miner who had survived rang his wife to tell her he was all right. What about Michael next door, she said. 'Don't ask', was the reply. She felt an 'unbelievable sadness' and then, a 'guilty feeling, because I still had a husband and she [her neighbour] didn't' (Background Briefing, 25/4/95).

Anger came later, when details of what had happened began to emerge.

The mechanisms that led to the explosion are well understood. Coal has a propensity to heat up slowly, when exposed to air. Unless it is well ventilated its temperature will rise, leading to what is called a 'heating'. The process is also called 'spontaneous combustion'. Eventually, the temperature may reach the ignition point of methane. If there is a significant concentration of methane in the area it will explode. This is what happened at Moura in 1994. The very same circumstances—a spontaneous combustion and a methane gas build-up—were responsible for the explosions at Kianga, near Moura, in 1975, in which thirteen men died, and at Box Flat in southern Queensland in 1972, when seventeen miners were killed.

There are plenty of warning signs as a heating develops: it gives off both a tarry smell and abnormal amounts of CO (carbon monoxide). Both these signs were observed in the weeks prior to the explosion at Moura, but for reasons to be explored in detail later they were ignored or dismissed.

A mine is divided into areas called 'panels', which are mined in turn. The day before the explosion, work was completed in one panel and, as was normal practice, this mined-out network of tunnels was sealed off by constructing a number of walls (seals) at relevant tunnel intersections. But what was not normal is that the sealing was carried out as a matter of urgency, in the hope that this would contain the threat of spontaneous combustion which, by that stage, management suspected might be occurring in the panel. Sealing meant that the area was no longer ventilated and the concentration of methane gas would rise, entering what is known as the explosive range at a predictable time. Despite this conjunction of factors, the men went underground on the night of the explosion to work in other parts of the mine, unaware of the danger they were in. As it turned out, the panel exploded just minutes after entering the explosive range, killing all in the vicinity.

A mine official later told the inquiry that he had felt no need to explain the situation to the men who went underground that night since they would have learnt about it via 'the grapevine'. It was this comment more than any other that angered the residents of Moura. As the daughter of one of the dead miners asked: 'Why did BHP expect my father to learn [about the danger] through the grapevine?' (Background Briefing, 25/4/95)

This anger comes through very clearly in an interview with a miner's widow, seven months pregnant at the time of the explosion:

> I go out there [the mine site] every couple of days.
>
> *Why's that?*

> Got nowhere else to go.
>
> *And what do you do?*
>
> We've got a little place out there where we can take flowers. I should say a lump in the dirt where we can take flowers, cause that's all we've bloody got. (Background Briefing 25/4/95)

The inquiry's ultimate explanation for what had happened was 'management neglect'. Here are its words.

> It is the opinion of the Inquiry that events at Moura surrounding assumptions as to the state of knowledge of the night shift on 7 August and the safety of those at the mine represents a passage of management neglect and non-decision which must never be repeated in the coal mining industry. Mineworkers place their trust in management and have the right to expect management to take responsible decisions in respect to their safety. They also have the right to expect management to keep them informed on any matter likely to affect their safety and welfare (Windridge 1996:41).

But members of the inquiry panel were insistent that this neglect did not amount to negligence. The inquiry chair issued the following instruction at one point: 'I really do not want to hear that word [negligence] again in these proceedings. It is a distraction which we do not need and it is not a consideration in these proceedings.' On another occasion, during cross-examination, a member of the inquiry panel made an off-the-cuff suggestion that perhaps some people had been negligent. He subsequently withdrew it, apologised for any offence he might have given, and substituted the word 'complacency'. The witness he was questioning was happy to accept that there may have been some complacency involved, but not negligence.

This remarkable sensitivity to the word 'negligence' alerts us to something important. The words 'neglect' and 'negligence' have very similar meanings in everyday English; indeed, they are treated as identical in section 75 of the *Queensland Coal Mining Act*. But to many lawyers, negligence implies individual culpability and the inquiry panel was keen to avoid attributing blame to individuals. It saw its aim as establishing the facts of what happened in order to make recommendations about how to avoid future explosions (Windridge 1996a:1). Had the inquiry been willing to entertain the idea that an *organisation* might be negligent, and hence culpable, it needn't have avoided the issue of negligence in the way it did. But given the panel's equation of negligence and individual culpability, the insistence of its chair that he was not interested in negligence can best be understood as an insistence on finding explanations that went beyond the individual. Unfortunately, the inquiry's failure to consider the possibility of negligence, either individual or organisational, left many Moura residents with a sense that justice had not been done.

This book is quite explicitly an organisational analysis, aimed at identifying organisational failures. In the long run this is more useful than any inquiry into the role of individuals. But the view taken here is that an organisational analysis does not preclude findings of fault. There are areas of law (e.g. Robens-inspired occupational health and safety acts) where organisations can be found to be negligent and hence subject to penalties. From this point of view an organisational analysis aimed at understanding what went wrong can sit side by side with a concern that justice be done. It is simply that an organisational analysis leads in the direction of a finding of organisational rather than individual culpability. Whether BHP as an organisation was indeed culpable is a matter on which views vary and on which readers will no doubt form their own view. In saying this I do not wish to imply that this is a matter of no consequence. Had the inquiry panel been able and willing to conclude that BHP as an organisation was culpable, this would have emphasised BHP's responsibility and vindicated the feelings of many Moura residents. Such a vindication would have helped make it possible to move on from the anger and perhaps to some extent the pain; it would have contributed to the process of healing that the people of Moura so desperately craved. Here is how one Moura resident described that craving.

> You pick up the pieces of your life and you stuff 'em back into a container of normality and you try to stuff 'em back into a cocoon of mundaneness and hope that one day it will all be together again. (Background Briefing, 25/4/95)

THE ARGUMENT

Very briefly, the argument of this book is as follows. The explosion was the result of a series of organisational failures: an inadequate system for transmitting information which meant that warning signs were ignored; a culture that denied the possibility of explosion, the failure of the organisation to specify who was responsible, which meant that no one in fact took responsibility; the absence of any incentives for managers to focus on the possibility of disaster; and lack of focus on the problem by the operating company, BHP. In short, the organisation from top to bottom was systematically inattentive to the potential for disaster.

To overcome this problem of organisational inattentiveness, organisations with the potential for disaster need to develop detailed plans for the management of identifiable catastrophic hazards. The plans must specify how relevant information is to be collected and dealt with and they must structure the decision-making process so that certain kinds of preventive action become mandatory under certain circumstances. Moreover, ways must be found to provide site managers with an incentive to implement these plans conscientiously, and company head offices must be held accountable for any failures in this regard. These points are spelt out in a little more detail in the following chapter outline.

Outline of book

Chapter 2 examines the literature on disasters in order to identify organisational factors of potential relevance to Moura. It shows how, with the exception of crude accounts which see disasters as inevitable, all the literature in one way or another focuses on management or organisational failures as the key to understanding these events. An important strand of writing identifies these failures as stemming from the priority of production over safety. Whether this was the case at Moura is one of the concerns of this book.

Chapter 3 examines the role of communication failure in the explosion. It argues that ways must be found to make information more visible to people in authority and that feedback mechanisms must be put in place so that those who provide the information have some indication that their input is being taken seriously. It suggests, moreover, that all this is a practical possibility, given the computer technology that is now available.

Chapter 4 examines aspects of the culture at Moura and shows how this served to nullify the early warnings of danger. First, there was a hierarchy of knowledge that prioritised personal experience and oral communication and thus undermined the effectiveness of written reports. Second, the organisation was paralysed by a series of beliefs, a culture of denial, which prevented recognition that a heating could be occurring. These cultural factors highlight the need for safety management plans that will overcome the tendency to organisational paralysis. These plans must define trigger events—that is, matters that require a response; they must specify what that response is, and they must identify who has the responsibility to take this mandatory action. The plans must find ways to make those who are specified as responsible actually accountable, and, finally, they must be effectively audited to ensure they are working in practice as described on paper.

Chapter 5 looks at the failure of management to exercise any responsibility for the safety of the men on the fatal night. It shows how, one after another, managers at various levels passed the responsibility downwards until it was ultimately up to the miners themselves to decide whether to go underground that night, even though they were in no way equipped to make this decision. These events demonstrate how important it is that organisations specify who is responsible for critical decisions.

Chapter 6 examines the process of auditing—that is, of checking that safety management systems are working as intended. It shows that auditing at Moura was hopelessly ineffective and highlights the importance of auditing if any confidence is to be placed in the effectiveness of safety management plans.

Chapter 7 raises the question of whether Moura mine management was serious about safety. It argues that the wrong measures of safety were being used and these systematically diverted attention from the control of catastrophic risk. This was a classic case of management attending to what was being measured and ignoring what was not.

More specifically, the mine measured safety by the number of lost-time injuries—that is, injuries that result in time off work. These can be described as high frequency/low severity matters. The mine had made great strides in reducing such injuries in the years prior to the explosion. But there was no attempt to measure the success with which management was controlling low frequency/high severity risks, such as explosions. The result was that in these matters management was, to say the least, complacent. This situation had come about because BHP had provided financial incentives to managers to reduce their lost-time injury frequency rate, but had provided no corresponding incentives to deal with rare but catastrophic events. The chapter recommends a change in the incentive structures for site managers.

Chapter 8 considers whether production took precedence over safety and concludes that, in relation to low frequency/high severity matters, it did.

Chapter 9 argues that part of the explanation for the disaster was that the operating company, BHP, had decentralised responsibility for safety to the level of mine management. This meant that the company exercised no effective oversight over how well mines were managing catastrophic risk. Most BHP mines have never had an explosion; BHP as a company has had several. It is therefore appropriate that BHP itself should develop the expertise to manage these risks and that responsibility in this area should lie with the company, not at mine level.

Chapter 10 takes up the issue of production versus safety again but in a broader context. Safety professionals frequently seek to motivate managers to attend to safety by arguing that 'safety pays'. This chapter shows that, in many cases and, quite surprisingly, even in the case of disasters, it is not an effective argument. It is ineffective largely because, although for the company as a whole it might be true that safety is profitable, there may be no corresponding advantage for the managers who are actually making safety-relevant decisions. In the Moura case, in particular, it seems that safety did not pay in any financial sense for crucial decision makers. Of course the emotional costs of disaster are enormous for managers who are implicated. It follows that those seeking to motivate managers to attend to safety would do better to emphasise the emotional rather than the financial costs of disaster.

Chapter 11 addresses the question of what type of regulatory system is best for the coal mining industry. It argues that debates about self-regulation versus prescriptive regulation largely miss the point. The real issue is whether or not inspectorates have the resources and the political backing to enforce whatever regulations they administer. This will be best achieved by moving coal mine inspectorates from their subordinate position in mining departments and locating them within generalist health and safety inspectorates which are in no way beholden to the coal industry. Furthermore, it is important that senior company executives be made legally accountable for safety failures. This is probably the most effective way to concentrate their minds on questions of catastrophic hazard.

The final chapter revisits the perspectives on disaster discussed in Chapter 2 and discusses their relevance for understanding what went wrong at Moura. It also summarises the organisational failings identified and the lessons to be learnt.

RESEARCH NOTE

The Moura inquiry produced more than 5000 pages of transcript, and thousands more pages of documents were submitted as exhibits. It also produced an excellent 70 page report (Windridge 1996) and a separate coroner's report (Windridge 1996a). This is a rich body of material on which to draw. The present book is based on a reanalysis of the evidence. Page references to transcript evidence are not provided here but are available from the author on request. Interviews were also conducted with various mine and company officials, with union representatives and with members of the inspectorates. My thanks go to all who contributed in this way.

2

PERSPECTIVES ON DISASTER

Before we begin a detailed investigation of what went wrong at Moura it is useful to look at a variety of perspectives employed to understand such disasters. Of necessity this requires us to examine what a number of different writers have had to say on the subject. For those with a social scientific bent this discussion needs no justification. But for the more practically minded, those who wish to get straight to the heart of the matter of disaster prevention, the chapter may need justification. It is this. In order to prevent disasters we must first understand what causes them. Looking at what others have had to say about the causes of disaster will sensitise us to possible causal factors at work in the Moura case. As this chapter progresses, therefore, it will highlight a number of factors to be examined in later chapters for their relevance to the Moura explosion. We shall see, too, that the most useful perspectives identify organisational failings, as opposed to technical problems or environmental circumstances, as the cause of disaster.

A distinction is sometimes made between natural disasters and those to which humans have contributed—man-made disasters, as they were called in an important analysis written in the 1970s (Turner 1978). Of course, even the most natural of disasters—cyclones, for instance—are only disasters because of human factors such as the failure to build cyclone-proof buildings or failure to take precautions against flooding. Nevertheless, there is a class of disasters where human activity is far more central to any causal analysis. Turner talks about these as sociotechnical disasters, with 'social, organisational and technical processes interacting to produce the phenomenon to be studied' (1978:3). It is these disasters that are of interest here.

The word 'disaster' will not be closely defined. I use it interchangeably with the phrase 'major incident'. Roughly speaking I mean by disaster an event involving multiple fatalities, but just how many deaths it takes to make a disaster is not a matter worth debating. Sometimes, as in the case of accidents involving major environmental damage, it may be appropriate to speak of disaster even though there are no human fatalities.

Perspective 1: Inevitability

The first perspective sees sociotechnical disasters as akin to natural disasters in that they are unpredictable and essentially unavoidable. They occur because humans are operating in dangerous environments where accidents are bound to happen and where it is inevitable that people will be killed or injured from time to time no matter what precautions are taken. This was the dominant view in the early stages of industrialisation and, certainly, coal mine explosions in the nineteenth century were viewed in this way. It was a view that remained at the forefront of thinking until comparatively recently. A study of the construction of the Snowy Mountains Hydro-Electric Scheme in the 1950s shows that tunnelling was assumed to be a dangerous activity in which it was inevitable that men would be killed. The only question was how many men per mile of tunnel. The construction authority achieved a figure of less than one death per mile, regarded at the time as one of the best safety records for tunnel construction in the world (Robinson 1997).

Even more recently, during the development of the oilfield in the North Sea off the coast of Scotland in the 1970s, accidents were seen as an inevitable consequence of 'operations which take place at the very frontiers of technology and in adverse climatic conditions' (see Carson 1982:5). Indeed, workers who died were seen as sacrificing their lives, 'dying for the greater good', as one government official put it (quoted in Carson 1982:44). One is reminded of the famous statement Western Australian mine owner, Lang Hancock, is reported to have made about dying asbestos miners: 'Some people have to suffer so the majority can benefit from asbestos' (The *Age*, Melbourne, 15 June 1978).

Such thinking is less in evidence today. Safety specialists

repeat endlessly that no death is inevitable and that companies can realistically aim for zero fatalities. Major construction and tunnelling projects now take place without a single death (e.g. the Sydney Harbour Tunnel and Sydney's Glebe Island Bridge). For present purposes, however, the principal point to be made about the belief in inevitability is that it leads nowhere in terms of developing strategies for prevention.

PERSPECTIVE 2: PRODUCTION VERSUS SAFETY

A second perspective on accidents, and more specifically disasters, is that they result from production taking precedence over safety. This has been demonstrated in numerous disaster studies. One of the most trenchant is Glasbeek and Tucker's account of the Westray coal mine explosion in Canada in 1992 in which 26 miners were killed. The authors show how the mine operator systematically ignored safety in the interest of production and how the provincial authorities in Nova Scotia failed to deal with the flagrant safety violations that were occurring.

But this is not a simplistic story of goodies and baddies. It is a sophisticated analysis of how a government sought to create employment and to ensure the prosperity of its citizens by encouraging coal production. To do this it guaranteed to buy output from the mine and lent funds to the mine owner on very favourable terms. Even the federal government chipped in by guaranteeing a large private loan which the operator needed to proceed. The operator was under pressure to begin production by a certain date because it had contracted to sell coal to a power station under construction for the provincial government that would begin power generation on that date. The point is, then, that the provincial authorities had a vested interest in going easy on the enforcement of health and safety requirements in order to keep the operator in business and production on schedule. Glasbeek and Tucker argue that this situation is inevitable in economies that are reliant on the extraction of commodities such as coal or oil. Australia, it should be noted, has such an economy. They argue that this 'political economic context creates an environment in which the protection of workers from harm is only a minor consideration in the decision making and behaviour of government officials and private investors' (1993:14).

Carson (1982) provides a very similar analysis of the high rate of accidents in the development of oil production in the North

Sea in the 1970s. He shows how the overriding imperative was to develop the oilfield as quickly as possible so that Britain could achieve self-sufficiency in oil. This was the highest priority both of the companies concerned and of the government department that was responsible for the development of the oilfield and at the same time for the administration of health and safety legislation off shore. Safety, in short, was sacrificed to speed of development. Here again the story is not simply one of capitalist baddies endangering the lives of workers but of a whole economic system in which production had top priority.

Nor is it only profit-making enterprises that are prone to put economic considerations ahead of safety. In 1986 seven astronauts died when the space shuttle *Challenger* caught fire soon after launch (see Vaughan 1996). NASA, the National Aeronautical and Space Administration, had been subject to economic constraints just as any private sector organisation and had compromised safety in numerous subtle ways in order to maintain its launch schedule and to carry out its program within budget. While there was more to it than this (see Chapter 4) and while there may not have been any conscious trading-off of safety against production, economic constraints were the ultimate source of the inadequate attention that NASA paid to safety.

There is an interesting irony here. The foregoing accounts show that, too often, safety is ignored in the interest of production. Yet safety specialists, both in government regulatory agencies and in companies where hazardous conditions exist, regularly claim that safety and productivity are not incompatible, in fact, that safety is profitable. How can this be? Clearly, for this claim to be at all credible, it must be carefully circumscribed. For whom is safety profitable? Under what circumstances is safety profitable? These questions will be taken up at greater length in Chapter 10. But, however they are answered, the evidence presented here suggests that it is a highly plausible hypothesis, to be tested in detail later, that production pressures contributed to the Moura disaster. In so far as production pressures are indeed the root cause of disaster, prevention will depend on generating incentives for safety sufficiently powerful to counteract these pressures.

Perspective 3: Normal accidents

One of the better known perspectives on disaster causation is Perrow's 'normal accident' theory (1984). Perrow's analysis focuses on high-tech systems with catastrophic potential, such as nuclear power plants. He defines normal accidents, or system accidents (he uses the terms interchangeably), as involving the unanticipated interaction of multiple failures of various of the system components.

The disastrous fire on the Piper Alpha oil platform in the North Sea in 1988 in which 167 men died was a system accident according to this definition (Appleton 1994). A maintenance failure led to a leak of condensate which exploded, igniting a large oil fire. The fire extinguishing system made use of electrically operated water pumps which failed because the initial explosion destroyed the power supply. Backup diesel-driven pumps that should have started automatically did not because they had been placed on manual start. They were on manual start to avoid any possibility that divers who were in the sea at the time might be sucked into the seawater intakes. The backup pumps could not be started manually because the oil fire made them impossible to reach. Even if it had been possible to start the pumps it is unlikely that the water deluge system would have worked effectively because the nozzles were partly blocked with rust. There were numerous other subsystem failures that contributed to the final outcome but enough has been said to indicate that this was truly a 'system' accident.

According to Perrow, such accidents can also be described as 'normal', in the sense that, though rare, they are virtually inevitable. There are two aspects of systems such as Piper Alpha that make accidents virtually inevitable: they are complex and they are tightly coupled. By 'complex', Perrow means that components of the system can interact in unexpected ways (1984:78). Thus the fire on Piper Alpha was fuelled by oil being pumped to the platform from adjacent platforms. No one had considered the need to interrupt this flow since it was assumed that any fire on Piper Alpha would be immediately dealt with by that platform's fire extinguishing system. By 'tightly coupled' Perrow means that the subsystem failures occur so quickly that there is not sufficient time for people to analyse what is happening and intervene in a way that might avert disaster. This was certainly the case on Piper Alpha.

What are the implications of this analysis for disaster prevention? Obviously, if systems could be made less complex or more loosely coupled this would increase the chances of averting disaster. But Perrow does not have much to say about how this can be done. His main point is that some of the most risky systems, such as nuclear power plants, are inherently complex, tightly coupled systems. In these circumstances the only way to prevent catastrophe is to abandon the activity altogether. For such a theoretically elaborate analysis this is a disappointingly mundane conclusion. Indeed, Perrow's analysis ends up as no more than a restatement of the inevitability perspective discussed above.

Furthermore, he may actually be wrong about the inevitability of disaster in the nuclear power industry. Since the Three Mile Island nuclear power plant failure in 1979, which came so very close to catastrophe, the nuclear power industry in the United States has recognised that an uncontrolled release of radiation at any one plant will mean the end of the entire industry in that country. Enormous effort has thus gone into guaranteeing reactor safety and the industry is demonstrably safer than it was in 1979 (Rees 1994). Perrow's work, based largely on the circumstances of the nuclear industry at the time of Three Mile Island, does not seem to have allowed for this development.

A further limitation of his analysis is that many of the disasters in which we might be interested are not normal or system accidents as he has defined them and therefore cannot be analysed in terms of system complexity or coupling. The gas leak from a chemical plant that killed thousands at Bhopal in India, the fiery destruction of the space shuttle *Challenger*, the Soviet nuclear reactor accident at Chernobyl from which people are still dying, the Exxon Valdez tanker oil spill in Alaska, an undoubted environmental disaster if not a human one—none of these is a normal or system accident, according to Perrow (1994:218). They are no more than 'component failure accidents' which cannot be analysed in system terms. 'They are alarmingly banal examples of organisational elites not trying very hard,' he says (Perrow 1994:28). Again, this is hardly a very helpful analysis from an accident-prevention point of view.

Perrow's analysis is even more disappointing in the case of mine disasters. These are on the whole not normal or system accidents, he says; most are simply component failure accidents. Moreover, mines are neither especially complex nor tightly coupled. His analysis, in the case of mines, again degenerates into the inevitability perspective: 'Mining is inherently difficult and dangerous' (1984:245), he says, and 'it would appear that there is an irreducible hazard in mining—an unpredictable environment for humans' (1984:251).

Interestingly, Perrow's pessimistic account of the long-run inevitability of 'normal accidents' in complex, tightly coupled systems has provoked a contrary school of thought known as high-reliability theory (HRT). This theory notes that many of the systems that Perrow describes—nuclear power stations, aircraft and airways—are in fact remarkably free from major accidents. HRT does not of course claim that high-reliability organisations (HROs) are entirely accident-free, merely that their accident rates are remarkably low. On the other hand, Perrow and his supporters do not claim that normal accidents occur frequently, just that they are bound to occur sooner or later. The debate between the two is thus about whether the glass is 99 per cent full or 1 per cent empty (Sagan 1993:48).

HRO theorists have, however, made some useful contributions to our understanding of why HROs are as reliable as they are. For one thing, HROs have 'nested authority structures' (LaPorte & Consolini 1991:31–2). This means that, while they routinely function in a highly bureaucratic way with rules and standard operating procedures designed to maximise operational predictability, at times of greatest pressure authority shifts downwards to frontline operators who are accorded considerable decision-making discretion. The rationale for this is that, at such times, senior managers, who are removed from the arena of operations, will not be able to respond as quickly as frontline operators; devolving authority at critical times increases the likelihood that decisions will be taken that avoid disasters.

Another finding is that HROs mobilise multiple observers of all critical events. The more eyes there are observing, the greater the chance that errors or warning signs will be noticed and reported. This insight will be developed at greater length in Chapter 3. Thus, while Perrow's normal accident theory seems almost devoid of insights about prevention, the school of theorising that developed in reaction to it, high-reliability theory, yields some potentially useful ideas.

PERSPECTIVE 4: MISINFORMATION

The curiously unsatisfactory nature of Perrow's analysis stems from the fact that it is based on a single case, the Three Mile Island incident. Given this genesis it is not surprising that the analysis fails to generalise readily or to offer insight into other types of disaster. Turner's work, *Man-made Disasters,* stands in stark contrast. Turner studied a wide range of disasters in an effort to identify factors common to all. This strategy enabled him to offer a far more generally applicable theory.

Stated in its crudest and most schematic form it is this: 'Disaster equals energy plus misinformation' (1978:189). It is a truism that disasters result from the inappropriate discharge of energy and this part of the theory needs no further discussion. More fruitful is Turner's insistence that all sociotechnical disasters involve an information or communication failure of some kind. The central question to ask then is: 'What stops people from acquiring and using appropriate advance warning information so that large-scale accidents and disasters are prevented?' (1978:195). Turner suggests a number of reasons why the relevant information may not be available.

1 The information is completely unknown. This is a very rare situation, particularly in relation to disasters this century. It is nearly always the case that information that might have averted disaster is available somewhere within the organisation concerned.
2 Information is noted but not fully appreciated. This may occur for a number of reasons, including a false sense of security that leads individuals to discount danger signs, pressure of work that diverts attention from warning signs, or difficulty in sifting the information from a mass of other irrelevant facts.
3 Prior information may not be correctly assembled. This may be because it is buried in other material, distributed among several organisations or distributed among different parties within the same organisation. The failure to assemble relevant information is a particularly common cause of disasters. On Piper Alpha information about maintenance work done by one shift was not passed on to the next shift because of inadequacies in the permit-to-work system. This was the start of the chain of events that led to disaster.
4 Failure to pass bad news up the line to senior managers. Turner includes this in the previous category but it is a sufficiently distinctive situation to warrant separate treatment. As Stone (1975:89) observes in a slightly different context:

> in literally every case of corporate wrong doing autopsied by myself and a group of University of Southern California law students, it turned out that *someone* down the corporate hierarchy was aware that trouble was brewing (usually for the company, in the last analysis, as well as for the public). For a variety of reasons the bad news never landed on the desk of someone who

> had both the authority and the inclination to do something about it.
>
> This is exemplified in the *Challenger* disaster. Information that earlier shuttle rockets had experienced technical problems was confined to one section of NASA and neutralised by a process of reinterpretation (see Chapter 4). Had this information been passed upwards to the highest-ranking NASA officials a different set of decisions would have been made (Vaughan 1996).

Turner's work provides a rich body of hypotheses about what might have gone wrong at Moura. The value of the approach is that it implies that disasters can be prevented by improving the flows of information and ensuring that relevant information does end up on the desk of those with the capacity and inclination to do something about it. A number of suggestions along these lines will be developed in Chapter 3.

PERSPECTIVE 5: LATENT ERRORS AND GENERAL FAILURE TYPES

One of the most useful perspectives on major accidents is that developed by James Reason (1990). His ideas are now routinely used in Australia by the Bureau of Air Safety Investigation in analysing aircraft crashes. Reason distinguishes between active and latent failures, as follows (quoted in BASI 1993:31).

> Active failures [are defined as] those errors or violations having an immediate adverse effect. These are generally associated with the activities of 'front line' operators: control room personnel, ships' crews, train drivers, signalmen, pilots, air traffic controllers, etc.
>
> Latent failures: these are decisions or actions, the damaging consequences of which may lie dormant for a long time, only becoming evident when they combine with local triggering factors (that is, active failures, technical faults, atypical system conditions, etc) to breach the system's defences. Their defining feature is that they were present within the system well before the onset of a recognisable accident sequence. They are most likely to be spawned by those whose activities are removed in both time and space from the direct human-machine interface: designers, high-level decision makers, regulators, managers and maintenance staff.

Reason argues that major accidents or near misses occur when

> latent failures, arising mainly in the managerial and organisational spheres, combine adversely with local triggering events (weather, location, etc) and with the active failures of individuals at the sharp end (errors and procedural violations).

He shows for instance that latent failures underlay Three Mile Island, *Challenger*, Bhopal, Chernobyl and Piper Alpha (1990:188ff). Thus his analysis is more generally applicable than Perrow's.

Reason goes on to argue that the purposes of accident prevention are best served by focusing on latent errors. Eliminating latent errors which may have given rise to an accident ensures that there will be no repetition. In contrast, seeking to deal with the errors of frontline operators by training, exhortation or threat is far less reliable. It is clear then that Reason is essentially drawing attention to the systemic or organisational causes of disasters as opposed to blaming individuals, as is so often done in these circumstances (see Hopkins 1995, Chap. 1).

Together with a number of co-workers, Reason has subsequently developed his analysis by identifying various types of latent errors (Hudson et al. 1994). These are called 'general failure types' (GFTs). After a great deal of empirical work the researchers have concluded that there are eleven GFTs, including such things as maintenance management, housekeeping, incompatible goals and communication. Communication is one of the more common general failure types. Thus, this analysis is consistent with Turner's theory that disasters or major accidents involve some element of misinformation. The theory of GFTs has given rise to the 'Tripod' system of accident analysis and prevention, now used by the Shell company worldwide (Hudson et al. 1994). The theory that disasters are ultimately due to latent or system failures has indeed proved a fruitful approach and the analysis of the Moura disaster in later chapters will be to some extent an inquiry into latent or systemic failures.

Perspective 6: The failure of organisational memory

A further perspective on major accidents begins by noting that they recur in frighteningly similar ways, very often within the same organisation. Moura 1994 was a classic case of this. It was the fourth underground coal mine explosion in Queensland in 22 years and the third to have been caused by spontaneous combustion. It was the third explosion in a BHP coal mine in Australia in fifteen years. The failure to learn from past events could hardly be more dramatic. (See Appendix 2 for a list of Australian mine disasters.)

Almost as dramatic was a fire on an Esso oil platform in Bass Strait in 1989, just nine months after the Piper Alpha fire. The circumstances of the two fires were remarkably similar, except that by great good luck the Bass Strait fire was extinguished and, although some workers suffered burns, no one died. The Victorian coroner who investigated the fire was so struck by the 'remarkable similarities in the respective causes of the North Sea disaster' and the Bass Strait fire that he devoted part of his report to demonstrating these similarities (Johnstone 1991:8). He further noted that in the nine years prior to his report there had been no fewer than twenty fires reported on Bass Strait platforms, at least one of them, in 1986, involving a fatality. It had exposed deficiencies in the fire-extinguishing system similar to those present in the 1989 fire. He comments that:

> a history of fires and incidents in the Bass Strait and overseas (including Piper Alpha) clearly indicates the obvious risks where even the smallest error can lead to a potential for disaster. The lessons ought to have been learnt well before 1989 (1991:12).

Why do organisations seem destined to repeat their own and others' mistakes? Kletz provides an answer of sorts in the title of his book, *Lessons from Disaster: How Organisations have no Memory and Accidents Recur*. The problem is, he says, that those involved on the first occasion move on or forget, and the organisation itself has no corporate memory. There is a convergence here with Turner's argument since the problem is essentially one of inadequate communication of information about past accidents. Kletz suggests a variety of ways to improve the corporate memory (1993:21–2) that involve the systematic reanalysis of old accidents and recommunication of the lessons to all concerned.

In particular, he suggests that managers need to undergo regular refresher courses on the kinds of things that can go wrong in their organisations and what should be done to prevent these occurrences. He makes the important point that it is not just a matter of training the workers on the job and their immediate supervisors. What is needed is training for managers at the highest level. This, he points out, is very seldom done. This analysis suggest that it is important to focus on the role of senior company management and their state of knowledge.

PERSPECTIVE 7: MANAGEMENT SYSTEM FAILURE

In recent years, following major accidents, various governments have taken action to prevent recurrence. Following a chemical discharge in Seveso, Italy, in 1976, in which hundreds were injured, the European Community issued the Seveso Directive; this requires all companies that handle large quantities of hazardous materials to demonstrate that they can handle them safely (Kletz 1994:88).

The requirement that organisations develop plans for the management of major hazards was also one of the recommendations that came out of the Piper Alpha inquiry. In the UK context, these plans are known as safety cases. Operators of hazardous installations must identify all the ways in which disasters might occur and plan both management systems and engineering strategies that will minimise the risk of such occurrences. These plans must be presented to the regulatory authority for approval. Safety cases have been credited with bringing about a substantial improvement in industrial safety in the UK (Cullen 1991:278).

The safety case strategy was adopted in 1993 for the management of safety in offshore oil installations in Australian waters (DPIE 1995; Barrell 1996). Most recently it has been picked up by Worksafe Australia in its national standard for the control of major hazard facilities (NOHSC 1996). Worksafe's initiative was triggered in part by the explosion in a chemical factory in St Peters, Sydney, in 1990 and a fire at a chemical tank farm at Coode Island, Melbourne, in 1991 (Emmett 1996:583). As we shall see, the Moura report adopted a similar approach, recommending that mines identify major hazards and develop safety management plans for each.

What all these responses have in common is that they see disasters as reflecting management system failures. The inquiry into Piper Alpha drew the following four lessons (Appleton 1994:182):

1 Management is responsible.
2 A systematic approach is required.
3 Quality of safety management is critical.
4 Auditing is vital.

These lessons are clearly all about rectifying the management system failures that led to the disaster. Interestingly, Kletz concurs with this approach in his most recent book, *Learning from Accidents.* He suggests that major accidents have 'layers' of causes, the ultimate layer being management system failure.

As a perspective, management system failure encompasses many of the previous perspectives. If communication is inadequate, this is a management problem; if latent failures are present, this is a management problem. It is appropriate therefore that governments have chosen to focus on this concept in their drive to prevent major accidents.

CONCLUSION

The analysis in later chapters of what went wrong at Moura will draw on several of these perspectives. In a sense I shall be hypothesising that the mechanisms identified in these accounts were operative at Moura. Where the data show this to have been the case, this will generate suggestions as to how Moura-type disasters can be avoided in the future. In large part this book will be an analysis of how the management system failed at Moura. The value of this will lie in highlighting aspects of management that must be carefully attended to and built in to safety management plans if disasters are to be averted. The issue of whether production took precedence over safety will also loom large, giving rise to suggestions about ways governments might deal with this problem. In the final chapter I return to these perspectives and identify their contribution to our understanding of what went wrong at Moura.

3

The Communication System

One of the recurrent findings in disaster research is that information that something was wrong was available somewhere within the organisation but was not communicated to relevant decision makers. Recall the very striking observation by Stone (1975:89):

> in literally every case of corporate wrong doing autopsied by myself and a group of University of Southern California law students, it turned out that *someone* down the corporate hierarchy was aware that trouble was brewing (usually for the company, in the last analysis, as well as for the public). For a variety of reasons the bad news never landed on the desk of someone who had both the authority and the inclination to do something about it.

This chapter outlines aspects of the *system* of communication at Moura that impeded the flow of safety-relevant information. It also makes recommendations about what can be done to remedy the defects identified. In the next chapter we examine aspects of the *culture* at Moura which further served to neutralise what warning signs there were.

Spontaneous combustion may occur in mines in such inaccessible places that it is not readily detected. It does give off a distinctive smell, often described as a 'tarry' or 'benzene' smell, but this may only be detected intermittently. It can also generate a visible heat haze. These signs were observed by mine officials from time to time in the weeks preceding the explosion. They were even recorded by deputies (work crew leaders—see glossary) in their end-of-shift reports on two occasions. But these reports were not effectively monitored by management and the warnings went unheeded.

A second set of signs was generated by the system of gas monitoring. Spontaneous combustion gives off carbon monoxide (CO) and the greater the rate of CO production the greater the cause for concern. The rule of thumb in the industry at the time was that a CO production rate of 10 litres per minute 'requires investigation', while a rate of 20 litres per minute indicates that 'considerable danger exists'. On 22 July, 16 days prior to the explosion, one of the deputies obtained a reading on his hand-held gas monitor which, after calculation, yielded a figure of 19 litres per minute. Given that the 10/20 figures are by no means exact, this was a very clear warning that considerable danger existed. The 19 litres per minute figure was discussed by a small group of mine officials and, for various reasons, judged to be an error. None of the decision makers apparently knew of the earlier reports of smells, which might have led them to view the reading in a different light. Neither the mine's registered manager nor his superior, the manager of the combined underground/open cut operation at Moura, knew of this reading (see Appendix 1 for management hierarchy). Most miners and deputies were also unaware of what had happened. In the week prior to the explosion, hand-held gas monitoring by deputies yielded further readings which, had the calculations been done, would have indicated a dangerously high CO production rate (Windridge 1996:32). But the calculations were not done and these warning signs went unrecognised.

Various witnesses testified that had they known about these signs prior to the explosion they would have concluded that a heating was occurring. They would then have taken decisive steps to deal with the problem and certainly not allowed men underground on the fateful night. One of the most remarkable facts uncovered by the inquiry is that, although there were numerous indications that a spontaneous combustion was occurring, the communication systems in operation in the mine did not effectively capture this information. As a result the organisation failed to respond appropriately.

THE SYSTEM OF WRITTEN REPORTS

Let us examine in more detail the communication system on which Moura relied, beginning with the system of written reporting. There were two written reporting systems in operation at Moura. First, deputies in charge of particular workplaces or work tasks filled out reports at the end of the shift. There might be anywhere from three to six such reports per shift. Second, the undermanager responsible for the whole shift filled out a report. He was generally expected to have read the deputies' reports and signed them before filling out his report.

This system might have been expected to result in the transfer of anything of particular significance from the deputies' reports to the undermanager's report. In fact it did not. The practice was that deputies' reports dealt with safety matters, as required by legislation, but shift undermanagers' reports dealt primarily with production matters—how much had been produced and whether there had been any impediment to production or down time. They did not cover safety matters. There was no legislative basis for this omission, or even any basis in formal mine policy; it was simply a practice that had evolved and which was passed on to new shift undermanagers by example. The result was the systematic exclusion of safety issues from the shift undermanager's reports. Thus, when a shift undermanager noted a tarry smell on 17 June, which he recognised as an indicator of an incipient heating, he did not mention this in his report. Again, when the shift undermanager noticed a smell and haze on 6 August and on this basis decided to recommend that the panel be sealed ahead of schedule as a 'precautionary measure', no mention was made of these observations in his report. Of course, shift undermanagers did need to communicate with their oncoming counterparts about safety from time to time, but this was done orally rather than in writing.

To summarise, Moura operated a two-tier written reporting system which confined safety to the lower tier. It was a system that in principle impeded the flow of safety-relevant information up the hierarchy or out to the widest possible audience.

The logical next question to ask is: Who read these various reports? An oncoming deputy was required to read and sign the report from the previous shift, although there is evidence that this did not always occur. Reports were also supposed to be read and signed by the deputy's own shift undermanager and the oncoming shift undermanager. The exception to this rule was at weekends when for much of the time there was no undermanager on duty. There was no procedure for ensuring that any manager read deputies' reports at weekends.

A critical instance of communication failure of this type occurred on the afternoon of Friday 24 June. On that shift a deputy noted a benzene smell and recorded this in his report. The oncoming deputy neither read nor signed this report. The evidence suggests that it was also unread by the next shift undermanager to come on duty. Quite apart from the situation at weekends there is evidence of considerable non-compliance by shift undermanagers with the requirement that they sign deputies' reports. In short, deputies' reports, which might contain vital safety information, were not reliably read by managers above them in the hierarchy, nor were they reliably read by other deputies. It should also be noted that there was no system for extracting relevant information from earlier reports. Thus, a safety matter that was missed at the time it was first reported effectively disappeared without trace under a pile of paper.

There is another tragic feature of this system. Deputies who filled out reports assumed that these reports would be taken seriously. They assumed that having noted a safety concern in their report their job was complete and others would consider the matter raised and take whatever action was necessary. There was nothing in this reporting system that required managers to report back to deputies about how they had responded to the matter raised. Deputies thus had no way of knowing that their reports might not have received the attention they deserved.

THE RELIANCE ON ORAL COMMUNICATION

So far we have been looking at defects in the system of written reporting at Moura. But there is a second aspect of the communication system that made it difficult to capture and assemble all the signs that a heating was occurring. At Moura, the primary means of conveying information of any sort was oral, not written. This was not just a practice; it was the policy, spelt out in the mines' Quality Assurance system. Thus, job descriptions for both deputies and shift undermanagers specified that they should liaise with their oncoming counterparts in order to communicate (orally) about matters of safety, among other things.

One of the consequences of this was that matters that were judged to be serious were handled in this way, by face-to-face communication; then, because the matter had been dealt with, they were not noted in written reports.

This assumption that information about important matters

would be conveyed orally had another unfortunate implication—namely, that matters about which only written reports were made were assumed to be not especially important. This further undermined the motivation for managers and deputies to pay careful attention to written reports. Indeed, one undermanager admitted that he did not read the reports of the deputies on his shift because he assumed they would have already communicated anything of importance to him orally.

Oral reporting is, of course, inherently unreliable. Just how unreliable became apparent at the inquiry when witnesses who claimed to have made oral reports to their superiors found their superiors denying that they had been told. Thus, for instance, the shift undermanager who noted the tarry smell on 17 June says in evidence that he notified the undermanager-in-charge. The latter denied this and claimed he had no knowledge of any reports of smells until just before the decision to seal. Whatever the truth of these matters it is clear that oral reports can be misunderstood and/or rapidly forgotten and are not accessible to anyone who might later seek to review the evidence.

Furthermore, practices at the mine were not always consistent with this principle of reliance on oral communication. The structure of shifts and the staffing arrangements at weekends meant that deputies did not overlap with each other; nor did the shift undermanagers who were on duty for some but not all the weekend shifts. Thus face-to-face communication between these people was limited at weekends. (Note that both the sealing of the panel and the explosion occurred at a weekend.) Evidence was given at the inquiry that even when shift undermanagers did overlap there was not a great deal of discussion between them.

In light of the policy of oral communication, one might have expected that miners coming on duty would be briefed at the beginning of the shift on what was happening. But there was no such briefing at Moura. Instead, miners were expected to brief each other during the 'hot seat changeover'. This was a policy whereby the outgoing shift would remain on the job underground until the new shift arrived to take over. This system of overlapping shifts was introduced in part in the belief that it would improve communication between shifts. But while hot seat changeovers meant that information about a particular task might be passed from one shift to the next, information about what was happening elsewhere in the mine, and other more general types of information, could not be effectively transmitted. In short, the hot seat changeover had serious shortcomings as a means of communication.

A further consequence of the primacy of oral communication is that managers who had been on leave relied on oral updates from other managers rather than reading through the reports written while they were on leave. Oral recall of this nature is of course highly selective and evidence given at the inquiry showed that managers who had been on leave were often unaware of events while they were away which might have led them to focus more carefully on the possibility of a heating.

COMMUNICATING DOWNWARDS—THE 'GRAPEVINE'

The preceding discussion has looked at why the communication system failed to capture information available at various points in the mine and channel it upwards to relevant decision makers. The system for passing information down the hierarchy was equally defective. Thus, for instance, following the high CO reading of 22 July, a decision was made to keep a closer eye on the matter by increasing the frequency with which deputies would take hand-held CO readings. This decision was not communicated to miners in general and nor was any rationale for the decision provided, even to the deputies who were expected to take the readings. It might also be noted that management's failure to communicate with the men about what was happening was a missed opportunity to ask them to be on the lookout for any further signs of heating—a missed opportunity, in other words, to enlist their help and involve them in their own safety. Communication downwards was so poor that not even the mine's engineer, who had originally reported the high reading to the manager, knew that the readings were being taken daily rather than weekly. Had he been aware of these daily readings he might well have done the relevant calculations in the crucial last week before the explosion.

Again, when a decision was made to seal the panel earlier than originally intended, in what management described as a 'precautionary measure', none of this information was conveyed in any systematic way to miners or deputies. The undermanager-in-charge said later that he relied on the grapevine in this matter and simply assumed everyone knew what was happening and why. The inquiry showed however that this assumption was far from justified and very few of those who worked at the mine had any real idea of what was happening. Had the miners and deputies been better briefed, there seems little doubt that they would have objected to going underground on the night of the explosion. (This point is substantiated in later chapters.)

This reliance by management on the grapevine is strikingly similar to the system of communication prevailing on Piper Alpha. There, managers relied on tacit communication; they assumed that people would 'just know' what to do without being told (Wright 1994:88). As it turned out, they didn't. There is even reference in the Piper Alpha report to information on Piper Alpha being communicated 'on the grapevine' (Cullen 1990:232). Not surprisingly, communication failure was found to be one of the factors that contributed to that disaster.

The Moura report spoke of the 'abysmal state of communications at Moura' (Windridge 1996:32) and on the evidence presented above this description could hardly be denied. But we should not necessarily draw the conclusion that Moura was exceptional or even unusual in this respect. As already noted, communication failure is routinely found to have contributed to disasters and corporate wrongdoing of various sorts. It is important therefore to draw lessons and to think about solutions that will be applicable not just to coal mines faced with the possibility of spontaneous combustion but to all disaster-prone organisations.

Learning the lessons: written communication

Two general lessons can be drawn from the Moura case on how to improve systems of written communication. First, a decent system of communication must find ways to make important information highly visible, so as to improve the chances that relevant decision makers and, more importantly, people who understand the significance of the information will become aware of it. The system of written reports by deputies at Moura failed dismally in this respect. Second, a decent communication system must have carefully thought through feedback mechanisms so that those who make the initial reports know what has happened as a result of their reporting. No such feedback occurred at Moura.

Visibility

The importance of making critical information highly visible has been brought out in the studies of high reliability organisations (see Chapter 2). This work has demonstrated that organisations which achieve disaster-free performance, despite the potential for disaster, do so by mobilising many pairs of eyes at crucial times. The presence of many observers increases the likelihood that warning signs will be picked up and small but critical errors will be noticed.

Air traffic control at major airports provides a good example. When traffic volume is low, a single controller does the job. As the traffic picks up, an associate controller assists. At high tempo

> a third, often senior, controller joins the two regulars as 'another pair of eyes'. At top tempo . . . the area supervisor . . . may also be nearby, along with perhaps two or three other controllers who are interested spectators (LaPorte & Consoline 1991:37).

American aircraft carriers provide a second, vivid example. Operations often require a dozen or more aircraft to land in quick succession, perhaps at 60-second intervals. The aircraft must be caught by an arrester wire and brought to a rapid halt; should the wire fail to capture a plane it must still have sufficient speed to get airborne again. Landings are therefore high-speed and very tense events. Despite the obvious dangers, these landings are almost entirely without incident. Part of the reason for this extraordinary safety record is the scrutiny of 'dozens of watchful eyes intently scanning the deck for any potential misadventure' (LaPorte & Consoline 1991:38). Moreover,

> as each aircraft comes round to the final approach, a television camera picks it up for video recording and later debriefing. These images are sent round the ship and into the squadron spaces on a hundred screens. . . . Meanwhile, ten pairs of eyes scan the deck to ensure that no newly parked aircraft, towing equipment or crew members stray over the designated 'foul lines' on either side of the landing area (LaPorte & Consoline 1991:38–9).

One interesting feature of this account is that the eyes need not be observing directly; it is possible to use modern technology (in this case, television) to increase the number of observers. I shall return to this point shortly. For the moment we should note that there were people within the BHP organisation who were particularly knowledgeable about spontaneous combustion. Had some of the warning signs at Moura been communicated to these people they would probably have reacted more decisively than management at Moura did.

Feedback

The second lesson in communication emerging from the Moura case concerns the need for effective feedback mechanisms. It is not enough that people make reports or pass information up the line. The outcome must be fed back to the person who made the initial report. This will improve the morale of reporters and they will be motivated to take the reporting process more seriously. In the absence of such feedback, reporting systems are likely to break down. Either reports will simply not be made or, where they are required, as in the case of coal mine deputies, they may degenerate into an almost pointless ritual.

Feedback is important not only to ensure that reporters take the process seriously, but also to obligate those to whom they report to act on reports conscientiously. Many companies now have reporting systems that specify the action to be taken and the person responsible for taking the action. The best companies attempt to 'close the loop' by requiring the person who is responsible for taking an action to 'sign off' when he or she has taken it. But even this may not be enough to ensure that the action taken is effective. The principle of feedback requires that the person who initially raised the concern be notified of what action has been taken.

To be truly effective the process must not terminate at this point. The next step is to require the person who initially raised the matter to indicate whether the action taken is satisfactory. Where the initiator is not satisfied, the matter should cycle through the system again until such time as the initiator is satisfied or, alternatively, a senior manager of the company is prepared to override the concerns of the initiator, in writing.

Such a system at Moura would have ensured that reports about tarry smells were taken seriously and that the possibility that a heating was in progress was responded to properly. Consider the fate of the deputy's report of a benzene smell on 24 June. We have already seen that both the shift undermanager and the oncoming deputy claim not to have seen this report. Nevertheless, the ventilation officer knew about it and apparently checked on the smell himself at the location where it had been reported. He could smell the chemicals that are routinely used to anchor roof bolts, but no benzene smell. As he said in evidence,

> When Reece [the deputy] said he smelt benzene, that's what I smelt [roof bolt anchor chemical] and I was not alarmed and I could tell the difference between benzene and chemical because, you know, I had smelt the lot before.

The ventilation officer, in other words, concluded that the deputy had confused the two smells and on this basis he dismissed the report. One inquiry witness testified that he had heard at the time 'on the grapevine' that the report had been dismissed in this way. This outcome was not, however, communicated to the deputy who made the original report; he apparently did not hear of it, even on the grapevine. He was very clear about the nature of the smell—he had smelt it before, he said, after a mine explosion and fire in 1986. It is likely, therefore, that had he been aware that his report had been dismissed on the basis that he had confused the two smells he would have objected. If his report had then cycled through the system again, as proposed above, it might have been given more credence and, when considered along with other indicators, it might have led to a somewhat different management response.

WRITTEN COMMUNICATION IN THE AGE OF COMPUTERS

With a reporting system like Moura's in which reports are written on forms and copies filed in one or two places, it is hard to imagine how the principle of visibility or even the principle of feedback could be effectively implemented. But with the widespread use of computers, electronically networked, it is relatively easy to design appropriate systems. I will always remember the chief executive officer of DuPont Australia showing me how, at a moment's notice, he could bring up on his screen a half-page summary of the most recent lost-time accident reported within the company, worldwide. The accident in question happened to be in one of DuPont's African plants.

The system of e-mail, in particular, has enormous potential. Shell Australia has devised an innovative system based on the e-mail technology. It is designed to collect information about health, safety and environmental problems that might be emerging anywhere in its vast network of petrol stations. Driveway attendants who are concerned about any such matters are invited to call a toll-free number and report it to a Shell telephonist. The telephonist types out the message which is then e-mailed to a variety of managers including OHS and environment managers at head office in Melbourne. Messages even appear on the screen of the general manager Melbourne, although there is a filtering process to ensure that he is not swamped. Messages are also archived in a database for later analysis to identify trends or emerging problems. At the time of interview I was told that there were five to ten messages coming through to head office each day in this way. Here, then, is a system that enables many eyes and, in particular, the eyes of relevant decision makers to view the concerns arising at the grassroots of the organisation. Its strength is that messages go far beyond the particular workplace where they originate and appear on the desks of people most able to take effective action.

There are, however, several problematic features of this system. First, mention was made above of a filtering system. If the number of messages is at all large, some filtering will be essential, otherwise messages will simply not be read. But filtering is a form of censorship and this has the potential to undermine the whole system unless it is very carefully designed. Perhaps what is needed is a person who can be relied upon to assess accurately the nature of the report and direct it to senior managers who have a particular responsibility for the matter. Second, Shell's system involves no provision for feedback to informants. In fact, in order to encourage people to come forward with reports, they are not even required to identify themselves. Third, at the time of interview Shell had not audited its system to see how well it was working. One driveway attendant I spoke with knew nothing of the system. At my suggestion he approached his manager for the relevant telephone number. He was told not to worry himself about such matters! He was however given a number to call in the event of EFTPOS difficulties! I have no idea how widespread this kind of subversion of the system is. Perhaps in the case of the Shell system it doesn't matter, since Shell is already getting a steady stream of reports that may be enough to identify the trends or issues in which the company is interested. But, in principle, any company that relies on such a system needs to audit it rigorously to ensure that all relevant reports are indeed being fed into the system. I shall deal more fully with this question of auditing in a later chapter.

Another way in which communication in organisations can be improved is by the use of computerised databases. These have the enormous advantage that they can be computer searched and thus matters of interest can be rapidly retrieved. They also have the capacity to highlight matters where action or feedback is required. Australian mines have already begun to make use of such databases.

Two questions have to be answered in developing such databases: what is to be entered and who is authorised to make entries. BHP's only currently operating underground mine in Queensland, Crinum, has what it calls a 'corrective action' database. Essentially, management, in consultation with the workforce, decides what corrective actions are needed and enters these into the system. The database indicates who is responsible for taking the action, when it is supposed to be completed and when it is actually completed. Regular reports are printed which enable tasks not yet completed to be identified. ARCO's Gordonstone mine has a similar procedure, but has augmented it with an e-mail system which prompts people who are overdue on their tasks. BHP's open-cut mine at Norwich Park has a very sophisticated computer system which allows rank and file workers to enter messages. If supervisors do not respond to these messages within a specified period, the messages automatically 'escalate' to a higher level. The process of escalation continues until a response is forthcoming. The system naturally generates data on supervisors who are not answering messages (Apostolou 1997).

These systems are invaluable. But not all safety-related information is captured in this way. What is needed in underground mining is nothing less than the computerisation of the system of reporting by deputies and undermanagers. Once such a system was in operation printouts of information about particular types of incident, critical or otherwise, could easily be arranged. In particular, as soon as there was the slightest suspicion that a heating might be under way, such a database could be searched for instances of smells, high CO readings and other signs that had been observed at the mine in recent weeks or months.

Such a system answers the two questions above very broadly. All safety-relevant and production-relevant information could be included and all shift undermanagers and deputies would be expected to fill out their reports directly onto a computer. There is no reason in principle why reports by the miners themselves could not be included in the system.

All these databases should be made as widely available as possible. Government inspectors, for instance, could have access to them. Moreover, senior management and safety specialists in company head office, possibly thousands of kilometres away, could have instant access and be able to search from their own desks for whatever information they wished. They might discover things that the people on site had missed. In addition, by assembling data from all their mines they might be in a position to identify problems not readily apparent at any one site. While such a system does not automatically ensure that many eyes are watching, as e-mail, in principle, does, it does mean that any eyes who wish to watch can do so.

It is important to reiterate, finally, that where computerised systems are based on reports made by individuals there must be provision for feedback to the individual concerned. As noted above, the reporter must be asked whether he or she is satisfied with the response given to the initial report. If not, the matter should be dealt with again at a higher level. In the case of deputies and undermanagers, because they are required to make reports routinely, they should be asked to indicate which aspects of their reports, if any, require a response. This would automatically direct the attention of their superiors to critical matters in these reports. Such a feedback system would ensure that those who received reports were far more accountable for their responses than is normally the case. (Further details of how this might be done for deputies' reports are available from the author.)

LEARNING THE LESSONS: ORAL COMMUNICATION

Where an organisation relies on oral communication, as Moura did, systems must be designed to ensure that it can and does occur. Hot seat changeovers are only a partial solution. At Crinum, located in a remote part of central Queensland, all deputies are provided with two-way radios in their cars. As they make the long drive to work, the outgoing shift undermanager gives them a detailed report on his shift. Oncoming deputies can, of course, ask questions and make comments. When they arrive at work they go to a conference room for briefing on the plans for the next 24 hours. Jobs are assigned, questions asked and plans explained. One can well imagine that, had such briefings been held at Moura, when the decision was made to seal the mined-out section ahead of schedule, questions would have been asked and the danger of allowing men underground while the section went through the explosive range would have been recognised.

Other types of meetings are also important. At Crinum, shift undermanagers meet regularly as a group. Moreover, production crews are briefed by their deputies underground at the beginning of every shift. They also have a weekly above-ground meeting. At Gordonstone such weekly above-ground production crew meetings are seen as a central element of the mine's safety strategy and may last for up to four hours.

In short, for effective oral communication to occur, an extensive and cross-cutting network of meetings must be carefully constructed. Matters cannot be left to the grapevine, as they were at Moura.

Conclusion

As every disaster analysis shows, safety depends on good communication. Furthermore, an effective communication system will be both oral and written.

Day-to-day communication depends on the oral transfer of information. This transfer must be carefully structured; it cannot be simply assumed that an overlapping shift system will achieve the objective.

The written component of the communication system is essential if organisations are to be able to keep track of themselves—that is, to learn from past events, to verify that things that were supposed to have been done were indeed done, and so on. The Moura experience shows that systems that rely on writing things down on paper are of limited value. Fortunately, there is another way. The new information technology means that information once recorded can be retrieved, organised, highlighted and scrutinised as never before. Although the new technology is widely available, the hard work of devising systems to make effective use of it is only beginning. If the potential can be realised, communication within organisations will be far more effective than in the past and disaster-prone organisations will be correspondingly safer. This all depends, of course, on scrupulous auditing of whatever system is put in place and on the provision of incentives for people to use the system conscientiously. These matters are taken up in later chapters.

4

CULTURAL IMPEDIMENTS TO RECOGNISING DANGER: THE NEED TO STRUCTURE DECISION MAKING

There were plenty of warning signs that something was wrong at Moura. In the last chapter we saw how these signs tended to get lost or passed over, in part because of failures in the mine's communication system. But there is more to it than this. The fact is that signs *were* noted from time to time. However, the mine had not laid down any procedures to be followed when these signs were detected. In the event the responses by mine officials were thoroughly inadequate and Moura suffered from an almost total organisational paralysis.

This is best exemplified by the response to warning signs provided by the system of CO (carbon monoxide) monitoring. As noted in Chapter 3, the greater the rate at which CO is given off, the greater the cause for concern. The widely accepted rule of thumb in the industry at the time was that a CO production rate of 10 litres per minute 'requires investigation', while a rate of 20 litres per minute indicates that 'considerable danger exists'. But what should happen when these thresholds are crossed? What kind of 'investigation' is required and what should be done when there is 'considerable danger'? Moura mine policy was silent on these issues. Mine officials were left to decide these things for themselves. For a variety of reasons to be discussed below, although the CO production rate was well above 10 and at times nudging 20, mine officials exercised their decision-making discretion by choosing to do almost nothing.

This chapter concentrates on the issue of CO monitoring and, initially, describes three ways in which the mine failed to respond effectively in this area. The bulk of the chapter is an attempt to understand why officials responded so ineffectively; I argue that a series of cultural assumptions served to nullify the warnings. Finally, I draw out the policy implications of the analysis.

SOME FURTHER BACKGROUND

At this point we need a little more background information. The area in which the explosion occurred was known as 512 panel, an area about 400 metres long and 100 metres wide. The height of the panel was 4.5 metres—the thickness of the coal seam. Development of this panel began in November 1993. This involved cutting a rectangular network of tunnels to the extremity of the panel. By the end of April this was completed and the extraction phase could commence. During this phase, miners retreat to their starting point, removing as much of the remaining coal as they can without danger to themselves. The mined-out area, called the 'goaf', is subject to roof falls and is a no-go area for the miners. Conditions in the goaf generate the most significant risk of spontaneous combustion. Hence it is during the extraction phase that the CO production rate must be carefully monitored.

The principal method on which Moura relied for CO monitoring was a bundle of tubes bringing air from various points in the mine to a single surface location where gas analysis was carried out automatically. This system provided measurements of each of the main mine gases at each location at intervals of approximately 12 minutes. It was virtually a continuous monitoring system. It had been introduced following the last explosion in the Queensland coalfields in 1986 at the adjacent Moura No. 4 mine. The system measured CO concentration, in parts per million. With certain additional information about airflows in the mine this could be converted to a CO production rate in litres per minute. This calculation was done manually, once a week, using the weekly average of the CO concentration figures.

At the beginning of the extraction phase, CO was being produced at a weekly average rate of about 2 litres per minute. This rose steadily until, on 10 June, the weekly average reached 11 litres per minute. In so doing it had crossed the first threshold. It continued to rise in an almost linear fashion after that and on 17 July, three weeks before the explosion, it was calculated at 14.6 litres per minute. But despite having entered the range requiring investigation and despite the relentless rise towards the 'considerable danger' level, nothing effective was done. This failure to respond when the weekly trend figures crossed the 'requires investigation' threshold is the first of the three ineffective responses I alluded to.

Apart from this continuous gas monitoring, deputies routinely took measurements using hand-held instruments, once a week. On 22 July one of the deputies took a reading which, on a calculation, yielded the alarming figure of 19 litres per minute. (Remember: 20 litres per minute indicates that 'considerable danger' exists.) Some hours later a group of mine officials went down to the location where the reading had been taken and took readings with other instruments. The earlier reading could not be replicated and they concluded that it was wrong and should be dismissed. This was the second inadequate response.

The continuous gas monitoring system was equipped with an alarm which was triggered to go off when the CO concentration reached a predetermined level, initially 7 parts per million. This it did about a week before the explosion. But the mine had no policy on what to do when such an alarm was triggered. In particular, no one calculated the rate of CO production in litres per minute, and so no conclusions could be drawn as to whether the rate was approaching the 'considerable danger' level. There was thus a tendency to regard the alarm as false.

In the absence of any policy on how to respond to alarms, one option was simply to raise the value at which the alarm was triggered, so as to reduce the frequency of 'false alarms'. In the week prior to the explosion the trigger appears to have been raised to a level of 8 parts per million. At this new level the alarm was breached several times. This corresponded to a CO production rate of roughly 18 litres per minute although again this calculation was not made at the time. Each time the alarm was breached it was apparently 'acknowledged' and cancelled without any consideration of what it might mean or whether further action might be appropriate. Mine policy said nothing about who was entitled to acknowledge alarms and the identity of those who did so remains unknown.

The result was that the value of the alarm system which had been set in place to provide an early warning of spontaneous combustion was completely undermined. This is the third instance of ineffective responses by mine staff to warning signs.

How might management have responded?

Before beginning the analysis of why the mine's response to the various warning signs was so inadequate it is useful to have some idea of how the mine management *might* have responded. Consider first the kind of investigations that might have been undertaken once the 10 litres per minute threshold had been reached.

To start with, management might have reviewed as systematically as possible all the available evidence. They might, for example, have read back through all deputies' reports or, more simply, asked all deputies and undermanagers whether they had at any stage noticed any smells or heat hazes. At least two reports of a tarry or benzene smell would have been forthcoming—these were described in Chapter 3. Putting the evidence together in this way would have made it clear that something was amiss.

Once the available evidence had been reviewed, certain further investigations could have been carried out. The mine possessed an infra-red imager. This was a hand-held instrument which, when pointed at a spot, would give an indication of its temperature. This could have been used to search for the location of any suspected heating. It had last been used for this purpose in 1986, eight years earlier, and had since 'languished in a cupboard' (Windridge 1996:50). The mine had developed no policy to encourage its use. In fact, management permission was needed before it could be used. This amounted to a presumption against using it. The outcome was that the instrument was not used at any stage to investigate the possibility of a heating in 512 panel. The idea of using it simply did not occur to anyone.

An even more remarkable omission was the failure to make use of another piece of equipment, a gas chromatograph. This instrument had been bought in response to a recommendation of the inquiry into the 1986 Moura explosion. Located above ground, it could produce a detailed analysis of air samples fed into it. The simplest procedure was to take a 'bag sample', by hand, and bring it to the surface for analysis. Samples could also be drawn from the tube bundle.

To understand the particular value of this instrument we need a little more background. Spontaneous combustion of coal begins at quite low temperatures and proceeds slowly at first. In certain circumstances the rate of combustion increases and likewise the temperature of the coal. Coal that is undergoing spontaneous combustion does not have the potential to ignite a methane gas explosion until it reaches 595 degrees, the ignition point of methane.

Remarkably, the gas chromatograph can provide information about what temperature the coal has reached and thus gives a clear indication of how dangerous the situation has become. Coal undergoing spontaneous combustion gives off a number of different gases, quite apart from carbon monoxide. Moreover, the gases given off depend on the temperature it has reached. Thus, by identifying which gases are present in a sample taken from the mine atmosphere it is possible to deduce the temperature of the heating, even though its location has not been identified. In short, once management suspected a heating, use of the gas chromatograph could have revealed how advanced it was and, in particular, whether it had become an ignition source for methane.

Why was this invaluable instrument not used? A number of factors contributed directly. First, as with the infra-red imager, there was no mine policy encouraging its use or specifying circumstances in which it should be used. Second, there were only two or three people at the mine trained to use it; requests by various others to be trained in its use had been turned down. Third, the people who had been trained to use it were very low in the Moura mine hierarchy and therefore had no responsibility to initiate its use. Fourth, knowledge that the instrument could be used to determine the temperature of a heating was limited, even among management.

What makes this all the more remarkable is that the instrument was kept ready for use at all times. A particular mine employee was responsible for 'calibrating' the chromatograph every day to ensure it would give meaningful results. But this was no more than a pathetic ritual since, as this man told the inquiry, the instrument had been used only about four times in a year and a half. At no stage was it used to check for a heating in 512 panel prior to the explosion.

The investigative procedures described above would almost certainly have confirmed the presence of a heating at some stage. For completeness, let us sketch the options then open to mine management. Had the heating been located in an accessible area, one possibility might have been to dig it out. This strategy had been used with a heating at the Moura open-cut mine a few years earlier. If the heating was in some inaccessible part of the panel, it might have been extinguished with water. This was a possibility because the panel sloped gently downwards. Thus, as mining retreated, the mined-out areas could have been flooded. This would have allowed mining to continue.

Had this not been possible, a third option was to cease mining for a period and seal off the panel. Methane gas leaks continually from the coal seam into the mined-out areas and it is normally diluted and removed by the ventilation system. But once a panel is sealed and ventilation ceases, the percentage of methane in the panel atmosphere builds up, eventually to the point where there is not enough oxygen to sustain combustion. The mine atmosphere is then said to have become inert. An inert atmosphere would extinguish any heating. Once this had occurred it would be possible, weeks or months later, to reopen the panel and continue mining.

Needless to say, none of these options was considered in the weeks prior to the explosion since the possibility that a heating was occurring had not even been investigated.

Why was the response to warnings so inadequate?

The preceding discussion points to a number of factors that contributed to the failure to use invaluable pieces of detection equipment. But the broader question remains to be addressed of why the response to the various warning signs was so starkly inadequate.

There is an obvious temptation to blame individuals for failure to take their responsibilities seriously enough. At the end of the day that is a judgment some might wish to make. But such a judgment does not advance our understanding of why these individuals acted as they did. Such an understanding is necessary if we are to find ways of ensuring adequate responses in the future. In what follows I argue that there are two broad phenomena at work which account for these inadequacies—first, what I shall call the hierarchy of knowledge and, second, the tendency to dismiss or rationalise away unwanted evidence, a tendency so pervasive that it can reasonably be called a culture of denial.

THE HIERARCHY OF KNOWLEDGE: THE PRIMACY OF PERSONAL EXPERIENCE

It is convenient for present purposes to classify knowledge into three categories, based on how it is acquired: personal experience, oral communication and written communication. At Moura there was a clear hierarchy of knowledge. Knowledge based on personal experience was by far the most influential: it was both better remembered and judged to be more reliable than information gained in any other way. In turn, information acquired by word of mouth was more influential than information conveyed by the written word.

There is nothing surprising about any of this. It has long been recognised that personal experience is the best teacher and good teaching technique is based on this principle. Nor is the fact that oral communication is more influential than written at all new. Research has shown, for example, that even 'top level executives prefer face-to-face interaction and even hearsay and gossip to analysis of factual reports' (Mintzberg, quoted in Clarke 1992:36).

I find myself quite dispirited by this phenomenon. Here I am, dear reader, trying to communicate with you by use of the written word, the least effective means of instruction. But despite its drawbacks, writing is my only chance of reaching you, so I continue.

It is important to understand how this hierarchy of knowledge operated at Moura, for it goes some way to explaining the responses we have observed. When the reading of 19 litres per minute was obtained on 22 July and orally communicated to more senior staff, the response was not to check the records to see whether there had been any other instances of high CO readings or tell-tale smells reported (written communication). Nor was any attempt made to ask other mine officials directly about whether they had anything to report (oral communication). Rather, the response was to 'check' on the matter personally, to see for oneself, by doing another measurement. When this failed to replicate the earlier finding, doubt was cast on that finding. The experience of those doing the checking was seen to conflict with the report they had received, and their own experience was judged to be more reliable.

Again, when a deputy reported a 'benzene' smell on 24 June, both orally and in writing, the ventilation officer's response was to go and check. He failed to detect the smell himself and, giving priority to his own experience, judged the earlier report to have been in error.

A particularly dramatic example of the importance of personal experience occurred around the time the decision was made to seal the panel. On the afternoon of Friday 5 August a deputy reported to his undermanager, both orally and in writing, that he had smelt a strong benzene smell which he recognised from an earlier fire at Moura No. 4 as the smell of burning coal. The undermanager acknowledged that this was an experienced deputy but told the inquiry that the report surprised him: he was not aware of any other reports of smells, he said. Given the communication failures discussed in Chapter 3, this was perhaps to be expected. 'I just couldn't believe that we would be getting a tarry smell there . . .', he said, because the CO levels had not risen. Moreover, he himself had smelt nothing that afternoon. He concluded that the deputy had been mistaken and may have detected the smell of grease drums in the area rather than burning coal.

The next day, however, the undermanager himself detected a smell and a haze. It was this personal experience that led him to recommend to the undermanager-in-charge that the panel be sealed. He told two of the deputies: 'We were going to be sealing 512 [panel] due to a haze that *I* had noticed—slight haze that *I* had noticed and goaf smell from *my* inspection' [emphasis added]. He further explained to the inquiry that, when asked at the time by another miner why the decision to seal had been made: 'I told him that there had been a change of conditions from the time I had been there on Friday afternoon shift . . . there was a slight haze and a goaf smell was evident . . .'.

This is a truly extraordinary statement. The previous day a deputy had reported a strong smell, undetected by the undermanager. But when the undermanager detects it the next day he describes it as 'a change of conditions'. It is evident to anyone willing to learn from the experience of others that conditions had not changed. What had changed was this man's experience. In his mind the experience of others simply didn't count.

Giving priority to personal experience was not just an unconscious tendency; it was a conscious policy. One of the men involved in responding to the 22 July report of 19 litres per minute was the superintendent of the Moura mine rescue station. His testimony to the inquiry is worth quoting at some length. Note that one of his functions was to train miners in aspects of safety.

When you were at the mine and you were going to investigate this reading and so forth, you didn't consider it necessary obviously to browse back through deputies' reports or undermanagers' reports or things like that?

No

. . . You didn't really think it was necessary to properly investigating [sic] that you look at those things?

No

. . . and likewise, you didn't think it was necessary to your proper investigation that you conduct some cross-examination of people about the history of the panel?

No

And that, given your state of knowledge, was a perfectly appropriate and reasonable approach to take?

Yes

. . . Do I understand your background correctly when I suggest that you are one of the people . . . who believes in going and testing things yourself and trusting your own judgement?

There is no substitute for going down the pit and having a look.

And that is certainly a view that may have been conveyed by you in teaching to your mines rescue people?

I would say so, yes.

There is absolutely no substitute for getting down there yourself and checking it out for yourself?

That's right.

This philosophy proved to be one of the most serious safety hazards at Moura. Its flaw is obvious. Where warning signs are fleeting and intermittent, as is the case with many of the indicators of spontaneous combustion, it is likely that signs detected by one observer will not be present when a second goes down to 'check'. In these circumstances the result is that warnings are systematically discounted.

So far I have dealt with the supremacy of personal experience in the knowledge hierarchy. I shall not deal here with the priority of oral over written communication; this has already been treated at some length in Chapter 3. I shall, however, draw attention to this relationship at various points below. But one final observation is necessary here. The hierarchy of knowledge is to some extent a psychological phenomenon. But at Moura it was more than this; it involved a conscious set of beliefs about the weight to be given to different types of information. For this reason I have described it here as an aspect of the culture of Moura. It is the first of two cultural phenomena identified in this chapter as responsible for nullifying the warning signs. I turn now to the second, the culture of denial.

THE TENDENCY TO DISCOUNT UNWANTED EVIDENCE: THE CULTURE OF DENIAL

One of the findings from psychological research is that 'new evidence appears reliable and informative if it is consistent with one's initial beliefs; contrary evidence tends to be dismissed as unreliable, erroneous or unrepresentative' (Slovic et al. quoted in Clarke 1992:30). Psychology teaches us that where evidence conflicts with belief the individual is in a state of 'cognitive dissonance', an unpleasant state that must be resolved by adjusting either the belief or the evidence. Where the belief is strong, it is the evidence that is adjusted.

A famous example of cognitive dissonance and the reaction to it is the case of a sect in the United States which believed that on a certain day there would be an enormous flood and only true believers would be saved, by a flying saucer that would pick them up. The day came and went without either the flood or the flying saucer arriving. The sect leader then announced that God had decided to save the world at the last minute, as a reward to the faithful. In this way the evidence that appeared to disprove the sect's beliefs was adjusted so that the beliefs remained unchallenged (Festinger et al. 1956).

At Moura it was believed that there was no significant risk that spontaneous combustion might be occurring. Had this belief been effectively challenged it would have entailed the disruption of production, and quite probably the closure of the panel for an extended period of time. Given the production pressures under which the mine operated (addressed in a later chapter), this is clearly a situation in which we would expect to find a strong tendency to dismiss any contrary evidence. And so it turned out: an extensive set of beliefs at Moura had precisely this effect.

Incubation period: it can't happen to us

The first of these beliefs was that there was an incubation period between the time the coal was first exposed to the atmosphere and the possible onset of spontaneous combustion. This period was thought to be at least six months. Since the panel was to be extracted in less than six months, this gave rise to a presumption that there was no real possibility of spontaneous combustion during the life of the panel. According to the registered mine manager, the practice at Moura was to design panels so that they could be extracted within this presumed incubation period. This was the primary means of controlling the problem. He had learned this, he said, by word of mouth, in a dinner conversation with two more senior BHP executives at the time of his appointment.

The mine's engineer had also learned, by word of mouth, that the mine was protected by a six-month incubation period. The ultimate source for this view he understood to be the report into the Kianga explosion that occurred in 1975, a presumably impeccable source. But the Kianga report simply noted that the Kianga mine had planned its panel extraction on the basis of 'an assumed incubation period of six months'. It in no way endorsed this assumption. The Moura report went further, stating quite authoritatively that 'reliance on the concept of an incubation period is not an adequate defence . . .' (Windridge 1996:27). The belief that Moura was protected from spontaneous combustion by a six-month incubation period was therefore without foundation. Despite this, it gained currency in a process of face-to-face communication which gave it a validity in the minds of hearers that it did not deserve. It is easy to see why, armed with this belief, officials at Moura were predisposed to dismiss the evidence of heating.

A different mining method: 'normalising' the evidence

A second belief that played an important part in counteracting the evidence concerned the mining method being used in 512 panel. This was a new method that left considerable amounts of loose coal in the goaf. Loose coal gives off more CO, in the normal course of events. As extraction progressed, the quantity of loose coal left behind increased and the belief was that the rate of CO production, in litres per minute, could be expected to rise slowly and continuously. It was further presumed that this was not a matter of concern. The ventilation officer was even willing to estimate that the rate would rise by about 1 litre per minute each week. It should be noted, however, that according to an expert report presented to the inquiry there was no basis for believing that such a rate of increase was 'normal'; on the contrary, the upward trend should have been a matter of concern.

This estimate of the rate at which the figure could be expected to rise had an important consequence. The extraction phase was expected to last 12 weeks. This meant that, starting from a base of 2 litres per minute at the beginning of the extraction phase, the rate could be expected to increase to 14 litres per minute by the time extraction was completed. This reasoning effectively undermined the rule of thumb. Since the rate was expected to rise to 14 in the normal course of events, the first threshold of 10 litres per minute could be ignored with impunity.

We saw earlier that, when the extraction process still had three weeks to run, the rate had already reached 14.6 litres per minute. But this did not cause any reconsideration; the increase had been reasonably slow and steady and that was all that mattered.

The interesting thing about this belief is that it served to rationalise or normalise the high readings being obtained. This process by which adverse events are normalised has been noted before as an important precursor to disaster. Vaughan (1996) describes it in her discussion of the space shuttle, *Challenger*, which caught fire and plunged to earth in 1986 killing the seven astronauts on board. The integrity of the rockets depended on certain rubber seals, known as O-rings. It had been discovered on several previous launches that they did not perform as required at low temperatures. Indeed, they malfunctioned. Nevertheless they had not failed totally. Over time the malfunctioning was reconceptualised as normal and the risk of total failure came to be judged acceptably low. (Vaughan describes this as the 'normalisation of deviance'.) The temperature on launch day was colder than at previous launches. But the technical malfunction had been normalised. The launch was thus given the go-ahead. This time the seals failed totally with catastrophic results.

The exponential rise theory

The third belief that served to undermine the value of CO monitoring at Moura was in some respects a corollary of the second. A slow and steady rise was normal; what *was* of concern was a sudden or 'exponential' rise. This would be an unmistakable indicator of a heating. Thus, for instance, the undermanager-in-charge noted that he was unconcerned when the CO figure crossed the 10 litres per minute threshold because there was no 'sustained steep upward trend'.

The thinking here was totally confused. It was widely known that previous heatings had been accompanied by an 'exponential' rise in CO. But the problem is that once such a rise occurs it is already too late to control the situation. An exponential rise is not a warning of potential danger; it is an indicator that an extremely dangerous situation has already developed. Conversely, a heating can occur without any exponential rise; it did at Moura. Tragically, reliance on the exponential rise theory resulted in a systematic tendency to dismiss other early warning signs.

The onus of proof

A further belief that contributed very directly to the culture of denial concerns the 'onus of proof'. In a criminal trial, at least in theory, the onus is on the prosecution to prove beyond reasonable doubt that the accused is guilty. If the evidence cannot sustain this burden of proof, the accused must be presumed innocent.

Remarkably similar thinking was present at Moura; it was presumed that the mine was safe unless the evidence proved conclusively that it was not. Here are some examples. The first time a deputy reported a benzene smell, an obvious explanation was that it was caused by a heating. But because the ventilation officer had not himself detected the smell he was predisposed to doubt the original report. Another possible explanation was that the deputy had mistaken the smell of roof bolt anchor chemicals for the tell-tale smell of a heating. Since this possibility had not actually been ruled out (the deputy was never confronted with this alternative explanation), it was accepted in preference to the possibility that a heating was under way. An almost identical fate befell the smell report the day before the sealing. This time the alternative explanation was that the deputy had confused the smell of coal burning with the smell of grease drums. Again, the reporting deputy had no opportunity to rebut this possibility. And, again, the theory that a mistake had been made was accepted in preference to the theory that a heating was occurring.

The high CO reading recorded by the deputy on 22 July was dealt with in exactly the same way. A predisposition existed to doubt it because the investigating party could not replicate the finding. A series of alternative explanations for the high reading was then hypothesised. One possibility was that the instrument was faulty. A second was that the deputy had operated it incorrectly. A third was that the deputy had read it incorrectly. A fourth was that a vehicle had been in the vicinity at the time and CO emissions from its diesel engine had affected the result. All these possibilities occurred to the men who dealt with the report. Since none could be ruled out, the original report was judged to be in error.

The common pattern here is that an alternative explanation is provided that would account for the findings. Since this explanation cannot be dismissed, this casts doubt on whether the findings can be used as evidence of a heating. The doubt is then resolved in favour of the alternative explanation and the possibility that a heating is under way is dismissed.

A very similar mindset prevailed at the time of the *Challenger* launch. Management asked for certainty that the O-rings would not work. The engineers could not provide this certainty. It was therefore presumed that the O-rings would work. In this way the doubt was resolved in favour of a decision to launch (Vaughan 1992:244).

When a doubt is raised about an operation where people's lives are at stake, is it up to those who raise the doubt to establish that a danger exists, or is it up to those conducting the operation to prove that it is safe? Where does the onus of proof lie? All too often, it seems, risky operations are assumed to be safe unless it can be proved otherwise. That at least was the presumption at Moura.

Putting the indicators in context

A final belief prevailing at Moura was that very little reliance could be placed on any one indicator in isolation. Indicators only took on meaning when viewed in their overall context. This view was expressed time and again by witnesses at the inquiry. What this meant in practice was that a report of a smell only became an indicator of a heating if it was confirmed by other reports or if it was associated with an abnormally high CO reading. Likewise, a high CO reading could ultimately be ignored unless it was part of a rapid and sustained rise. Generally speaking, no single indicator could be interpreted as a sign that a heating was actually occurring unless it was confirmed by other indications.

While this view might appear reasonable at first sight, it was problematic for two reasons. First, it implies that unless warnings are confirmed they can in effect be ignored. This is another manifestation of the belief, discussed above, that the mine should be presumed safe from spontaneous combustion unless the evidence proved otherwise.

Second, if the significance of a warning sign depends on the context, then the appropriate response is to investigate the context thoroughly. If a smell is reported the obvious contextual question to ask is: How many other reports of smells have there been? This question was never asked. To argue that the significance of an indicator depended on the context but to make no systematic effort to establish the context nullified the effect of the warning.

Each of the beliefs discussed above dismissed the evidence in a different way: it was irrelevant, to be expected, inconclusive, unsubstantiated, out of context, erroneous or inappropriate. Given such an overwhelming culture of denial it was inevitable that officials would use their decision-making discretion, in effect, to ignore the warnings that had been given.

STRUCTURING DECISION MAKING

So far in this chapter we have explored two major sets of factors that contributed to Moura's organisational paralysis: the reliance on personal experience and the culture of denial. These two phenomena ensured that the officials at Moura would use the wide discretion they enjoyed to dismiss what evidence there was. What lesson can be drawn from this analysis? There is little hope of changing the disposition to rely on personal experience. As other research has shown, this is deeply ingrained in our ways of knowing. Nor does it make a lot of sense to try to change the various beliefs that go to make up the culture of denial. This was an extensive set of beliefs at Moura and, even though some of them might be corrected or changed, others would very likely develop in their place. Given that evidence of spontaneous combustion is so unwelcome, a culture of denial is almost inevitable.

The way around this problem is to limit the discretion exercised by decision makers or, to put the point more positively, to structure their decision making. It is only if companies develop policies to guide their managers in these matters and, of course, enforce these policies that the inherent tendencies towards paralysis can be overcome. Based on the Moura experience, these policies must, in particular, define the triggers to action, specify to some extent the actions required and identify who is responsible for taking the action.

Triggers

A triggering event is one that requires someone to take action. The 10/20 rule of thumb had two such potential triggers: when the rate of CO production reaches 10 litres per minute an investigation is required; if it reaches 20, urgent action is required. This could be contrasted with the other rule that was implicitly adhered to: where there is a rapid or exponential rise in CO an investigation is required. Apart from the fact that once a rapid rise occurs it is probably too late, the main problem with this rule is that it is ambiguous—what is a rapid rise? Since this is not clear-cut, it is always possible for a decision maker to argue that a rise is not rapid and therefore that nothing needs to be done. There is, in other words, no trigger that forces action to be taken.

Of course, triggers will often be arbitrary: should an investigation begin at 10 or 12 litres per minute? There is no rational way of choosing. But as long as triggers are set conservatively this is of little consequence; the aim is to ensure that decisive action will be taken before a crisis point is reached.

The alarms raised on the CO gas monitor represented an alternative trigger. Given that the alarm was going off at about 18 litres per minute, this was indicating not just the need for investigation but that considerable danger existed. The alarm was both visible and audible. But this very dramatic trigger to action was ignored because the mine had no effective policy on what to do when this alarm was given.

These triggers are quantitative; they specify a threshold value beyond which something must be done. Qualitative triggers are also possible. For instance, any report of a 'benzene' smell could be treated as a trigger requiring further action.

Action to be taken

Triggers must trigger action. In the case of a warning smell or gas reading, the action required could be to investigate whether there have been other warning signs and to collect further evidence by whatever other means might be available. Such a policy at Moura might have made it mandatory to employ the gas chromatograph and the infra-red imager. Moreover, policy needed to specify what should be done when alarms sounded on the continuous gas monitor. It was the absence of such a policy that allowed the alarms to be acknowledged and then ignored.

Who is to take action

It is no use specifying that action be taken without also identifying who must take it. At Moura, the ventilation officer, the engineer, the undermanager-in-charge, the mine manager and the safety/training manager all took some part in the decision making about the indicators of spontaneous combustion. Not one of them had any clearly specified responsibility to take action. A glaring example of this failure was that no one was designated to respond to gas alarms. Unless someone is charged with the responsibility to make decisions there is no guarantee that the right decisions will be made, or even that any decisions will be made.

Accountability

Decision makers must be accountable for their decisions. The best way to do this is to require them to commit to writing their decisions and reasons for them. At Moura, reports of smells were discounted because it was assumed that the reporter had made a mistake. If decision makers had been required to put this in writing they might not have jumped to such conclusions so readily.

A variation on this accountability requirement is to require that decisions to do nothing be justified in writing, while decisions to take further exploratory or preventive action need not be. In effect, this builds in a presumption that action should be taken unless the decision maker is prepared to provide written reasons to the contrary. Such a bias towards action might help overcome the biases towards inaction identified earlier in this chapter.

Auditing

Finally, and most importantly, any policy that adopts these principles must build in some mechanism to guarantee that the policy is working as intended. It must, in other words, be effectively audited. I take up this whole question of auditing in a later chapter.

Conclusion

Apart from the communication problems addressed in the last chapter, at least two other, cultural phenomena were at work undermining the effectiveness of the warning signs at Moura. One of these was the tendency to rely on personal experience in assessing the evidence; the other was an elaborate culture of denial. Together, they effectively paralysed the organisation. Similar cultural phenomena have been identified as contributing to other disasters.

Management decision making must be structured so as to counteract these phenomena and ensure that effective action is taken. Decision-making discretion must be limited by introducing imperatives to action that rule out the option of doing nothing. This is now increasingly recognised in disaster-prone industries. The Moura report recommended that mines develop safety management plans for each major hazard they face, and a good deal of thought has gone into developing guidelines to what should be included in these plans (NSWDMR July, October 1995). The principles that I have drawn from this analysis of what went wrong at Moura figure prominently in these guidelines. There is reason to think that they will be applicable in many industries where the potential for catastrophe exists.

5

THE FAILURE OF MANAGEMENT RESPONSIBILITY

One of the failures identified in the previous chapter was the absence of any clear specification of who was responsible for particular, safety-related decisions. All too often, nothing was done, or at least nothing effective, because no one had any specific responsibility to take action. One instance of this problem stands out above all others in the Moura story: the decision, or non-decision, to send men underground on the night the mine blew up. What happened on this occasion was nothing short of a fiasco. As the inquiry report noted it was 'a passage of management neglect and non-decision [making] which must never be repeated in the coal mining industry' (Windridge 1996:41).

This chapter focuses on how this fiasco occurred, for two reasons. First, it reveals a great deal about how responsibility for safety diffused downwards in the organisational hierarchy at Moura, to the point where it was very nearly a policy of every man for himself. Second, the decision of whether workers should be withdrawn because of imminent danger is one that may confront workers and managements in a wide range of industries. The lessons here are therefore of very broad relevance.

First, some more background. Mining the panel was finally completed on Friday 5 August. The intention at that stage was to seal the panel early in the next week. But on Saturday morning the shift undermanager contacted the undermanager-in-charge and persuaded him to authorise sealing immediately. The sealing was completed just after midnight on the Saturday night. The shift undermanager had been concerned about smells and a haze he had encountered and had urged that the panel be sealed early as a 'precautionary measure'. It should be noted that the decision to seal at the weekend was not without cost—workers had to be called in on overtime to do the job. During the Saturday there were further reports of smells. The undermanager-in-charge was informed of some of these warning signs. On the Saturday afternoon he did a calculation and discovered that the rate of CO production was 19 litres per minute. Remember that, according to the rule of thumb, 20 litres per minute indicated that considerable danger existed. When asked at the inquiry why he had done this calculation he replied that he was 'looking for something to do to try and give me some comfort'. It is clear that despite the denials of those involved, they 'suspected that there was a heating in progress within the 512 panel and that the panel was sealed with some urgency' (Windridge 1996:36).

Sealing the area meant that it would no longer be ventilated, and it was clearly understood that the concentration of methane in the sealed area would therefore rise, entering the explosive range some time after 11.30 on the Sunday night. Eventually, if the methane level continues to rise it moves beyond the explosive range because there is ultimately not enough oxygen to sustain a fire. This is what management hoped would happen. Of course, if a heating was occurring, the coal might well have reached a temperature high enough to ignite methane. Nevertheless, management appeared simply to hope in some confused way that the sealed area would pass through the explosive range without ignition (Windridge 1996:34). In the event the mine exploded at 11.35 p.m., five minutes after it was predicted to enter the explosive range.

Management was aware that the panel would enter the explosive range late on Sunday night. A new shift was due to go underground at about 10.30 p.m. The vital question for management was: Should these men be sent underground as scheduled or should all miners be kept out of the pit until the panel had passed safely through the explosive range? The panel could have been expected, all going well, to pass out of the explosive range within perhaps two days. Keeping the men out might thus have meant the loss of at most two days' coal production.

The decision was to send the men underground. Eleven miners began work in and around a production panel adjacent to the recently sealed 512. All died. A further ten, working in another part of the mine, managed to make their way out through dense clouds of dust.

Previous practice

Before we address the question of how the decision was made to send the men underground, we need to know something of past policy and practice at Moura. This mine, Moura No. 2, had been in operation for about 24 years and during that time 28 panels had been sealed. The great majority had been without event. The normal practice at the mine was for the men to go underground to work elsewhere in the mine when a recently sealed panel was going through the explosive range. In contrast, at the adjacent Moura No. 4 mine, the practice had been for panels to be sealed on a Friday and for the men to stay out of the mine over the weekend while the panel went through the explosive range. Moura Nos 2 and 4 were both BHP mines; so why was the policy different? The answer is that BHP management had no policy on the matter. At Moura No. 4 the miners had simply refused to go underground while a panel was explosive; at Moura No. 2, a different workforce had not taken collective action in this regard. In the absence of company or mine management policy on this issue it was apparently up to the miners to decide and implement policy for themselves. As we shall see, this was part of a pattern at Moura.

Although normal practice at Moura No. 2 was work as usual after a sealing, there were at least two occasions when miners had stayed out. The first of these was in April 1986, after 5 North panel was sealed. On this occasion CO readings from the panel had suddenly begun to rise very rapidly and there was no doubt in anyone's mind that a heating was occurring. It was obvious to all that the miners should stay out on this occasion.

The second was the sealing of 5 North West panel in 1991. On this occasion the levels of CO had gone from 9 to 12 litres per minute and the manager decided to seal the panel as a 'precautionary measure'. He did not believe that an ignition source was actually present at the time and he did not propose to withdraw the men. The miners and their deputies confronted the manager and demanded that the men be withdrawn while the panel was explosive. They were concerned, not only about the possibility of a heating but also about the possibility of a frictional ignition, a spark caused by a rock fall. According to one miner, the manager initially resisted and even threatened to cancel the deputies' tickets. However, after further discussion he agreed to the request and the men stayed out for two days.

Although the policy question of routinely withdrawing men

from the mine had not come to a head at Moura No. 2, it was a simmering industrial issue. From time to time men raised it with management and one deputy in particular had frequent discussions, even arguments, about the need for a policy of keeping the men out when a panel was explosive. The response he said he was given was that this was just an excuse to get time off and that it was impractical because 'it could take three months before the panel became inactive' (a substantial exaggeration). This deputy told the inquiry he had twice been suspended for arguing his case.

Management's decision

The question that management posed for itself that Sunday afternoon was not, should the men go down, but what should be done if the men expressed any concerns about going down. The difference is critical. Management saw this as an industrial relations matter, not a question of safety. When the shift undermanager raised it with the registered mine manager, who was not on site, his response was that no one should be forced down the mine and that the undermanager-in-charge should be contacted if the men expressed concern (see Appendix 1 for management hierarchy). In so saying, he effectively sidestepped any question of safety and removed himself from the decision-making process, handing his responsibility further down the line.

Early in the evening the shift undermanager phoned the undermanager-in-charge, also off site at the time, and raised the matter. His response was the same as his superior's, namely, that no one should be forced down the mine. He noted that there was plenty of alternative work for the men to do if they refused to go underground. But he went further than this. He told the shift undermanager not to raise the matter with the crews because if they had any concern they would raise it themselves.

In a very revealing interchange at the inquiry the shift undermanager was asked what would have happened if the men had been told that 512 panel was 'sealed last night due to a tar smell, a goaf smell and haze'? His response was that the men 'would have sat on their dig[nity]' and refused to go down. Given that the whole issue was conceptualised as an industrial relations matter, it is clear why management chose to tell the men as little as they could.

The question of safety was, however, touched on at one point.

The undermanager-in-charge asked the shift undermanager whether *he* felt it was safe to go down. The shift undermanager replied that it was so safe that he would go underground and 'kiss the seals' if necessary. Having satisfied himself on the safety question in this way, the undermanager-in-charge reverted to the industrial relations issue and told the shift undermanager that if the men refused to go down he should go down 'as matter of principle', in effect, to shame them. This was perhaps an unconscious attempt to play on miners' concerns not to be thought of as wimps, a point I shall develop shortly.

The question of whether the shift undermanager felt it was safe to go down appears to have been the only time the question of safety was explicitly raised in these deliberations. In asking this question the undermanager-in-charge effectively shifted responsibility for safety further down the line. Everything was made to hinge on his subordinate's assessment.

The shift undermanager is a relatively lowly official in the mine hierarchy (see Appendix 1). This particular man was asked at the inquiry whether he regarded himself as a professional. No, he said, he was 'just like a foreman'. He noted that, unlike mining engineers and other colliery professionals, he was never given the opportunity to attend conferences on mining matters. Moreover, he was singularly ill-equipped to shoulder responsibility for safety. He claims that he had never received any training at Moura about spontaneous combustion, that he was unaware of the significance of the CO rate in litres per minute and that he had never heard of the 10/20 litres per minute rule of thumb. He had not read the recommendations arising out of the inquiry into the 1986 Moura No. 4 explosion, even though he had been there at the time. In short, responsibility had been delegated downwards to a point where it could not be effectively exercised.

From management's point of view, then, the shift undermanager had been instructed not to withdraw the men but nor to force them underground if they did not wish to go. In effect, it was to be their decision whether they went underground that night. But lest they exploit this situation, neither the miners nor their deputies were told that the decision was theirs.

Decision making by the deputies

It is often the deputies who advise the rank and file miners in these situations. What role did they play in the decisions that were made or not made that night? It should be remembered that deputies are responsible for the safety of their own crews; they are not responsible for the overall safety of the mine. This affected their responses, as we shall see.

One deputy in the shift prior to the explosion was clearly concerned about the danger. During his shift he went to the surface on six or seven occasions to monitor the concentration of methane and the progress of the panel towards the explosive range. However, he recognised that the mine would not go explosive until the next shift and he felt under no obligation to involve himself in the decision that would then have to be made about whether the crew for that shift should go underground. In any case, because there was no shift overlap, there was no opportunity before he went home for him to communicate his concerns orally to any of the oncoming deputies.

Another deputy, also not on duty at the time but aware of some of the earlier indicators of spontaneous combustion, had made a definite decision that he would not take his specialised drilling crew underground while 512 panel was in the explosive range. Not only did he tell his crew of this decision, he told the undermanager-in-charge that he 'wouldn't take the blokes down while it went through the explosive range'. According to this man's testimony, the undermanager-in-charge made no reply.

What is interesting about this is that the undermanager-in-charge apparently did not question the deputy's authority to make this decision. The implications are staggering. Responsibility for safety had been fragmented to the point where it was conceivable that some crews might be kept out of the mine while for others it was business as usual.

A further view expressed by some deputies was that the men would, or should, be given the opportunity to decide for themselves whether to go down. As we have seen, management had already decided that the men would be allowed to make this decision but had not told them so. Here is how one deputy expressed himself at the inquiry.

> With hindsight, would you have expected management to have told you something about the conditions there on that night?
>
> *Yeah, yeah.*
>
> What would you have expected them to tell you with hindsight? What sort of things?
>
> *Well, I thought they would have told us that it was going through the range and it was up to you*—up to the individual as to whether he wanted to go or not *[emphasis added]*.

> What about a direction that nobody should go underground whilst it was in the explosive range? Would you have expected that?
>
> *That would be better still.*

In this interchange the deputy expresses the view that when the miners' lives are at stake it is appropriate that management give them the responsibility for deciding whether to go underground. This is an intriguing comment to which I shall return shortly.

What of the deputies who went underground that night? Did they make responsible and informed decisions about the safety of their men? Counsel for BHP at the inquiry believed so. Consider this interchange which took place with one of the two deputies who survived that night.

> And so you got information from his [another deputy's] report, and you also got information from the Unor [gas monitoring] screen about the levels of the gases and so forth, and you connected them in your mind as to the significance of the methane, so you had all that information?
>
> *Yeah.*
>
> But you nonetheless took the view that it was reasonable to go down the pit, didn't you?
>
> *Yeah.*
>
> You must have done, you went down?
>
> *Well I went down, hey?*
>
> Yes, and you are not a reckless person, are you? You are not safety reckless?
>
> *No.*
>
> You would pay due regard to proper safety questions?
>
> *Yeah.*
>
> And behave accordingly?
>
> *Yeah.*
>
> So, if you thought in truth that there was any problem about going down that night you would not have gone down?
>
> *That's right.*
>
> Now doesn't it follow from that, that in your assessment that night, taking into account what you had seen and learned and read, you nonetheless made the judgement that it was reasonable to go down?
>
> *Yeah.*
>
> And that's a judgement which, in your experience, would be reasonable for other people to make, wouldn't it?
>
> *Possibly, yeah.*

In this interchange, counsel persuades the deputy to agree

that he is well informed and responsible and that he had formed the considered view that it was reasonable to go underground that night. One would have to say, however, that these were not the spontaneously expressed opinions of this man; they were extracted from him in a line of questioning that allowed him no alternative.

Fortunately, this deputy gave a more realistic account at another point in proceedings. He had arrived at work later than the others and was surprised to find that they had gone underground. He expected that no one would be sent underground that night. He had presumed that management would have kept the men out. It should be noted that he was not alone in this presumption: various other deputies, not on duty that night, had also assumed that management would make a decision to keep the men out, given that the panel had been sealed as a matter of urgency.

But to return to our deputy: since the men, including two other deputies, had already gone underground, he presumed that a decision had been taken that the mine was safe. What went through his head was: 'Well, they must know something I don't know.' So down he went. At no stage did this man make a considered decision. In fact, he was not in a position to make a considered decision. He knew nothing about the previously detected indicators of spontaneous combustion and had no understanding of the significance of the data in litres per minute of CO. Not unreasonably, he relied on management to make the decision for him.

What of the second deputy who went down that night and survived to tell his story? Although he had a general belief that miners should not go underground when panels were explosive, he had no particular fears on this occasion. He had heard about the suspected heating and had been underground when smells had been detected the previous day. However, he had not detected any smells himself. He therefore wondered what the 'panic' was about and could not understand why the sealing had been brought forward. He regarded those who had made the decision as a 'pack of wimps'.

This man's comment alerts us to another factor operating in the minds of both deputies and miners which to some extent overrode any private fears they might have had: the culture of masculinity. Here is an interchange with this deputy.

If there was some concern that a miner had about safety, but it was

> not a serious concern, would you think that there might be some reluctance for a miner to mention that concern because he didn't want to appear to be perhaps a bit of a coward or a bit of a wimp, being worried about something which shouldn't be regarded as serious?
>
> *Yeah, I would say you could draw that conclusion.*
>
> Have you experienced that sort of thing yourself?
>
> *Yes, I have.*

A NOTE ON BHP BADGERING

The deputy just quoted was clearly a man whose judgment about safety-related matters was clouded by a concern not to be seen as a wimp. Yet counsel for BHP described him as a 'very, very experienced deputy and a very, very experienced Mines Rescue bloke', a man whose judgment could be trusted.

How can we account for the approach taken to these deputies by counsel for BHP? In particular, the line of questioning of the first deputy is so contrived that it calls for some explanation. Note, first, that this was not an isolated occurrence. Consider the predicament of another deputy in the following interchange.

> If you formed the view that there was a situation likely to cause danger, you had a mandatory obligation to stop work, suspend the operations until you had been able to go to the manager and get special instructions about it?
>
> *Yes . . .*
>
> You didn't at any time in relation to 512 suspend the work, did you?
>
> *No*
>
> Can I take it from that then, bearing in mind your mandatory responsibility, that you at no stage formed the view that there was any danger to the men?
>
> *I had a concern. Like I said yesterday, I don't put it in the context of danger, no.*
>
> You didn't at any stage think there was a danger to the men, did you?
>
> *Not on that shift, no.*
>
> Nor at any occasion of being in the 512 panel on the occasions you were?
>
> *No . . . I never formed that view, no.*

This man has been badgered into saying that, although he had a 'concern', at no stage did he ever think there was any 'danger' to the men.

Why is counsel for BHP so intent on getting deputies to conclude that in their considered view there was no danger to the men? According to counsel for the miners, this was all part of a 'protracted onslaught upon the humble rank and file miner and deputy to render them responsible for the destruction of the mine and the death of the men'.

A preferable view is that counsel for BHP was trying to establish that deputies were engaged in a responsible decision-making process and that these deputies had come to the judgment that 512 panel did not represent a source of danger. BHP counsel was clearly wishing to imply that if responsible and knowledgeable deputies could come to this view then it was not surprising that more senior managers were unaware of the danger. In so doing counsel acknowledges the point being developed here, namely, that at Moura, responsibility for safety-related decisions had been passed a long way down the organisational hierarchy.

Rank and file miners

Let us consider, finally, the decision-making processes of the rank and file miners who went underground on the night of the explosion. There was no general discussion among them about the situation in 512 panel before they began work that night. Nor were they briefed in any way about what had been happening. The undermanager-in-charge had explicitly said to the shift undermanager that such a briefing would be unnecessary, because the men were 'well aware of the situation as it was'.

He was quite wrong in this, however, as is apparent from the testimony of those who survived. One miner who went underground that night did not even know that 512 panel had been sealed until after the explosion! Another knew that the panel had been sealed but did not know that it was going into the explosive range. Most knew that the panel had been sealed because of a heating but had discovered this only from general talk among themselves; they had not been told by management. Moreover, they had very little idea of what the significance of a heating was. One told the inquiry he did not understand that a heating could cause an ignition, and that in any case he had heard it was 'just a small heating'. Another view was that once the panel had been sealed there was no further danger.

A computer screen with the results of the continuous gas

monitoring was available for miners and deputies to check at all times. Some of these miners looked at the screen before they went down but most had no idea what they were looking at. They looked at it just because others did!

The state of mind of one of these miners is well captured in the following interchange.

> Were you interested to know what was happening in relation to the build-up of gases in 512?
>
> *I don't know. I never even thought about it actually. I don't know . . .*
>
> Were you interested to see whether 512 was moving into the explosive range?
>
> *I don't know really. I didn't even give it any thought that night, you know.*
>
> Well, you were about to go underground?
>
> *Yes*
>
> Were you concerned at all?
>
> *I don't know. Not really . . .*
>
> Did you ever entertain the possibility of [an explosion] happening in 512?
>
> *No, it never crossed my mind, no.*

In fact for these miners it was just business as usual. Here is an interchange with another miner.

> What was your own view as to whether there was any danger in you going underground?
>
> *Well, my own view was we went to work on the Sunday night to do our normal work practice and it didn't concern me that we were going down the pit because that was our normal practice. We just went to work to go down the pit and cut coal and that was it.*
>
> Did you have any niggling concern in you mind about whether there could be an explosion in 512?
>
> *There could have been but in my mind it was safe to go down the pit because it was just normal for us to go down the pit at No. 2 when a section was sealed.*

This man assumed that if there was any danger he would be told. He certainly did not expect to be making a decision himself about whether it was safe to enter the mine.

Only one of the miners who went down that night, and survived, did consciously think about safety. He knew of the various reports of smells and he knew that the panel was entering the explosive range. He said to his mate: 'Maybe we shouldn't go down.' His mate replied that it was normal practice to go down when a panel was going through the explosive range. After that, he said, he 'never thought any more about it . . . we just jumped in the Rover and went down'.

To summarise, the rank and file miners who went underground that night had no awareness of the danger they were in. This is not surprising since they had been given no information by management about the prevailing circumstances. Their knowledge of spontaneous combustion was negligible and they had no way of understanding the meaning of the various indicators that were available. Even though many of them were aware from 'talk around the mine', as they put it, that a heating had occurred in 512, they were quite unaware of the danger it posed. These men were in no position to take decisions about their own safety and trusted that management would act responsibly in this matter. They had no idea that management had in effect given them the option of not going underground that night.

Expecting miners to risk their lives

The conclusion that follows from this analysis is that, at Moura, responsibility for safety-related decisions was diffused downwards to the point where in this case miners were 'tacitly expected' to make their own 'life and death decisions', to use the words of the inquiry report (Windridge 1996:41). But a stronger statement is possible. At Moura, and indeed more generally in the coal industry, there was a quite explicit assumption that it was reasonable to expose miners to danger, provided they voluntarily assumed the risk. From this point of view, the most reprehensible aspect of management behaviour on the night of the explosion was that the men were not given a clear indication that the decision whether to go underground was up to them.

Consider the circumstances of the sealing of Moura's 5 North panel in 1986. As noted earlier this was a panel in which the rate of CO production had taken off 'exponentially', indicating to everyone that a heating was far advanced and that an undeniably dangerous situation existed.

One of the deputies involved in assessing the situation had survived the Kianga explosion in 1975 and was very aware of the similarities between what had happened there and what was now happening in 5 North panel. He telephoned a government inspector who confirmed the urgency of the situation and ordered sealing as a top priority. Management expressed concern about getting the machinery out before the panel was sealed but was told by the inspector that it was far too late for this. 'Stuff the machinery', he is reported to have said. 'If they wanted to save the machinery they should have sealed off a week ago.'

The manager then called in men for the job of sealing and 'told them that it was a dangerous situation and they did not have to go down if they didn't want to'. The deputy talked to the men and told them his view was that the situation wasn't quite as advanced as Kianga. He advised them to go down to finish sealing. Relying on his advice, the men decided to go down.

The deputy, a union official and the government inspector were all present while the sealing work was carried out. The deputy admits that both he and the union official were 'starting to panic' about the situation. The inspector 'never actually panicked but I could see the perspiration coming down his forehead and I knew he was worried . . .'. The three, including the inspector, worked alongside the men to finish the job as quickly as possible. While the deputy was working he 'kept thinking of Kianga'. At Kianga, not only had the CO readings escalated, but smoke was coming from the panel. Nevertheless, men had been sent in to seal it. The mine blew up as they worked, killing thirteen.

As it turned out, 5 North panel did not explode at the time it was being sealed, nor subsequently as it went through the explosive range. But the deputy's experience in 5 North panel coupled with his experience at Kianga led to an impassioned plea to the inquiry that it 'bring down some recommendations for future sealings of sections where men don't have to go down and risk their lives'. In particular, he asked for recommendations that panels be sealed long before the levels of CO indicate that a heating is well advanced.

What is noteworthy in this account is that management appeared to have absolved itself from all responsibility for sending men underground by telling them that it was dangerous and giving them the opportunity not to go. Here we see, quite explicitly in operation, the belief that miners are ultimately responsible for their own safety. What is all the more interesting about this is that it was not just a management belief; it was widely shared. A union official and even a government inspector thought that it was acceptable for men to risk their lives in this way in an effort to prevent the mine from exploding. The men had made a choice and in so doing had accepted responsibility for whatever might befall them.

In times of war people knowingly risk their lives for their country. It is noble to do so: death in these circumstances is regarded as a sacrifice that the individual has made for the greater good. The same mentality is evident in the mining industry, except that, here, men are expected to risk their lives for the sake of the mine. The pressure on miners to take this risk is, of course, enormous, for if the mine is lost their jobs are in jeopardy. In effect, then, the expectation is that miners should risk their lives to secure their jobs.

The inquiry into the Kianga explosion was horrified by the way management had assumed it was acceptable to put the lives of miners at risk in order to secure the mine. It made the following recommendation: 'The prime consideration in the control of any heating or fire must be the safety of the personnel present.' But this recommendation fell on deaf ears: safety was clearly not the prime consideration in the 1986 sealing of 5 North panel.

CONCLUSIONS

There are various conclusions that may be drawn. Let us start with the most general—the question of responsibility. Among the many management failures revealed in this account, the one that stands out above all others is the fact that no one had specific responsibility for safety-related decisions. The critical safety question of whether the miners should be sent underground when the panel went explosive was sidestepped first by the mine manager and then by his subordinate, the undermanager-in-charge. Everything was made to depend on the opinion of a self-described foreman who was clearly not equipped to make this decision and for whom the relevant test was whether he felt safe enough to 'kiss the seals'.

The lesson is that, when it comes to questions of safety, management must have policies that specify precisely who is responsible for what decisions. These individuals must be held accountable for their decisions so that they apply their minds as conscientiously as possible. One way of doing this is to require them to commit to writing their decisions and the reasons for them. For particular decisions it may be possible to specify what factors the decision maker must take into account. These written statements should be routinely forwarded to superiors, inspectors and any interested parties so as to enhance accountability. This is all part of the wider policy of structuring the decision-making process more tightly (discussed in the previous chapter). My examination in this chapter of a particular decision, the most critical in the lead-up to the disaster, suggests that the need to specify who is responsible for what is one of the most important aspects of this structuring process. It is for this reason that safety management plans must pay particular attention to identifying decision makers and holding them accountable.

Furthermore, the more critical the decision, the higher should be the level at which the decision is taken. Decisions about sealing panels and about allowing miners underground while panel atmospheres are explosive are so critical they should be made at the highest possible level, and certainly no lower than the registered mine manager. I shall argue in a later chapter that these are really matters in which more senior company management should be involved. The argument will be that they should be seen as matters of company policy, and not left to the discretion of individual mines.

A final conclusion concerns the presumption that it is appropriate for management to give workers a choice about whether they work in dangerous circumstances. It is not enough that workers have a *right* to stop work when they fear for their safety. As the Moura story makes clear, there must be an *obligation* or *duty* placed on employers to withdraw workers from danger.

Modern occupational health and safety legislation places an overarching responsibility on employers to maintain a safe system of work. Employers have a 'duty of care' to their workforce in this respect. Inviting workers individually to assess the risks and volunteer to carry out dangerous work is not consistent with this duty of care. Coal mine safety acts (at the time of writing) have no such overarching duty of care and it is this omission that has allowed attitudes to persist that hold workers responsible for placing themselves in danger.

6

AUDITING

The term 'audit', as it is most commonly used, refers to a process of checking that organisations are managing their financial affairs competently and honestly. In safety circles the term has come to refer to the process of checking on the adequacy of an organisation's procedures and performance in relation to safety. There are numerous kinds of safety audits, including technical and machinery checks, compliance audits (i.e. checking that organisations are complying with relevant rules and procedures, both their own and government-imposed) through to the most general evaluations of the system by which safety is managed.

Auditing is one of the most widely used (and abused) ideas in the area of safety management today. It covers anything from a ten-minute exercise ticking boxes on a questionnaire, done by an administrative assistant whose boss is too busy to do it, to a three-week inquiry by a team of six high-powered managers from company sites or headquarters in other parts of the world.

Just as financial auditing provides no guarantee against fraud, so safety auditing to date provides no guarantees against disaster. All too often, inquiries following disasters reveal that safety audits that had been carried out failed to identify the problems that led to the disaster.

Consider again the Piper Alpha oil platform which was consumed by fire with the loss of 167 lives in 1988. Inadequacies in the 'permit to work system' meant that there was a lack of coordination between shifts. As a result, someone tried to operate a pump that was undergoing maintenance and started a fire.

Moreover, the water deluge system did not work, one of several problems with the system being that the nozzles were blocked with rust. The deficiencies in the permit to work system and in the fire control system should have been picked up in audits, but weren't. As one of the inquiry panel noted (Appleton 1994:181):

> Compliance with the permit to work procedure was monitored each day and had been one subject of a parent company audit just six months before the disaster. No deficiencies were reported. An annual fire safety audit was undertaken but its report had not referred to the problem of blocking deluge heads.

He went on:

> Clearly there was no shortage of auditing of the Piper platform and the way it was being operated. What was deficient was the quality of that auditing. Not only were departures from laid-down procedures not picked up, but the absence of critical comment in audit reports lulled senior management into believing that all was well.

BHP AUDITING

The story at Moura was depressingly similar. Numerous audits had been carried out, the most thorough being the so-called 'corporate' audit conducted by BHP at Moura and several other mines in Queensland just six months before the explosion. The audit had involved three people from BHP's Brisbane office, plus personnel from the various mines. But although it was an in-depth exercise, coming up with a long list of suggested improvements, it overlooked completely the issue of spontaneous combustion.

BHP has since drawn two conclusions about the failure of this audit to identify any of the deficiencies that led to the explosion. The first was that the audit was not sufficiently high-level. To have people within one division of the company checking on each other compromised the process since they would be reluctant to criticise their peers (Seymour 1996). BHP corporate audits now involve people from parts of the BHP empire that are organisationally and geographically remote from the enterprise they are auditing. In a subsequent high-level audit of several Queensland operations the audit team consisted of five senior managers from overseas BHP sites plus one independent academic.

The other conclusion that BHP drew from the failure of its corporate audit was that it was not focused sufficiently on catastrophic events such as explosions. The subsequent high-level audit *was* focused in this way and found, among other things, serious problems at another underground mine, specifically in relation to the management of spontaneous combustion (BHPAC 1996:28).

THE GAS MONITORING AUDITS

Moura's gas monitoring system was audited twice prior to the explosion. Since this system was critical to the detection of spontaneous combustion, these audits are worth examining in some detail. The audits were carried out in 1990 and 1992 by an external, expert auditor and organised by the coal mine inspectorate. Both audit reports failed to identify the deficiencies in the system that were highlighted in the inquiry.

The 1990 report notes, for instance, that when a gas alarm occurs 'the alarm is accepted by the entering of the respondent's cap lamp number'. Each individual had a unique cap lamp number and the system was intended to make it possible to identify who it was that acknowledged the alarm. However, it was revealed in the inquiry that people acknowledged alarms by entering any randomly chosen pair of numbers, thus defeating the purpose of the system. The auditor had simply relied on what he was told and had not picked up this practice.

The system had two alarm levels. According to the auditor: 'At alarm level 1 any available personnel accepts the alarm and follows up with an underground supervisor (probably an undermanager). If alarm level 2 occurs the manager is contacted and mine emergency procedure comes into play.'

But again the auditor was simply reporting what he was told. The inquiry found that alarm level 1 had been breached on several occasions in the days prior to the explosion and acknowledged by unknown persons without any follow-up with an underground supervisor.

The auditor also made the following comments in his 1992 report. 'There was no written gas alarm emergency procedure available in the control room. The procedure was known by personnel but a written procedure avoids any mistakes or critical omissions with regard to an emergency response.' He suggested, too, that 'emergency trial alarms . . . should be carried out on a regular basis'. These were useful suggestions but they were never implemented, in part because the inspectorate failed to pass on the audit report to the mine's management!

Auditing of the gas monitoring system thus failed in two ways. First, it did not audit the crucial part of the system, namely, what happened when the alarms were raised. It simply reported what was supposed to happen. One way to have audited this aspect of the system properly would have been to study what happened last time the system had alarmed. If this was not possible, an alternative might have been to trigger an alarm in some way and observe the response. This failure to verify that the alarm system worked as intended seriously undermined the value of the audit.

The second way in which the process failed concerns the failure to pass on the audit recommendations to management. This in itself is a serious omission but it points to a systemic failure—namely, that there were no follow-up procedures to ensure that audit report recommendations were implemented. Any such follow-up would have revealed that the report had not been received by management. Without follow-up procedures the value of audits is again undermined.

Quality Assurance Auditing

Quality assurance (QA) is a way of assuring customers and other interested parties that an organisation's products or services are of the required quality. More precisely, it is a way of assuring others that the organisation has a system designed to achieve this end. Purchasing firms are increasingly requiring of their suppliers that they be quality-assured and the result is that large numbers of firms, worldwide, have structured their management systems so as to achieve QA accreditation. QA is essentially about documenting the way things are supposed to be done, and then auditing to ensure that they are being done as intended. Where 'non-conformances' are identified these are fed back so that organisations may take 'corrective action' and so 'continuously improve' their performance.

In so far as safety-relevant matters are documented, safety can be subject to QA auditing. But can safety auditing carried out within this framework be relied upon? Overseas evidence shows the answer is no. Consider the following QA failure (Department of Transport 1996). On 23 February 1995, an aircraft owned by British Midlands Airways had just begun a routine flight when the oil pressure in both engines dropped to zero and it was forced to carry out an emergency landing. Subsequent investigation showed that both engines had just undergone maintenance and that the oil sealing system had inadvertently not been refitted. Engine test runs, which would have revealed the problem, had not been done. How did this happen?

The airline had recently moved from a system of quality control to quality assurance. This meant that maintenance work was no longer examined at critical stages by inspectors before the next step could begin. Rather, those doing the work certified on paper that they had completed all the necessary tasks involved in the maintenance process. To implement this system, maintenance planning engineers had specified the tasks to be performed for each maintenance job and devised a corresponding set of task cards. Maintenance workers signed off on each card as the task was performed. This was the system. The airline's QA department had the job of making sure that this system worked. It did so by collecting the signed cards and ensuring that the paperwork was in order, so as to satisfy the requirements of the regulatory agency.

There were two other kinds of checking the QA department might have done, but didn't. The first was to do spot checks on work in progress to ensure that maintenance workers were doing what they said they were doing, and doing it properly. The second was to evaluate critically the task description cards designed by the maintenance planning engineers, to ensure that they represented a satisfactory set of procedures. In this case, although the aircraft manufacturer's maintenance manual specified that engines should be test run before the job was considered complete, this was not specifically required in the set of procedures devised by the airline's maintenance planning engineers. The result was, as subsequent investigation revealed, that in about 50 per cent of cases engines were not test run.

In summary, the QA auditing system in this case was largely confined to checking paperwork. It exhibited two crucial deficiencies: it failed to consider whether the system being audited was a good system, and it failed to verify whether the people concerned actually did what they said they did. As we shall see, this was also true of QA auditing at Moura.

Quality assurance at Moura

Moura mine was accredited in 1994 by the standards association of Australia as quality-assured under AS3902. Only certain of the mine's procedures were documented and assured in this way. AS3902 is not a safety assurance standard and safety procedures were not initially covered, although it was intended that they would be at a later date. In particular, the procedures for dealing with spontaneous combustion, for sealing panels and for monitoring CO production all remained undocumented and not subject to audit at the time of the explosion.

Although the mine's system of communication was not spelled out in any single document, communication was addressed in some of the documented procedures. Thus, one four-page document about underground procedures contained a series of statements about communication at the beginning and end of shifts. For instance: 'Prior to the start of each shift the oncoming undermanager shall communicate with the offgoing undermanager to obtain the following information.' There followed a list of things to be discussed. Moreover, 'hot seat changeover' was specified as follows: 'The mining crews shall then travel to their nominated mining sections where they should communicate with the offgoing mining crew . . .' Again, a list of topics follows. Clearly, the system of communication on which the mine relied for its safety was actually defined in considerable detail in this document.

So how effectively was this document audited? The answer is: very poorly. The mine's quality manager had done an audit but had not tried to audit the communication aspects, for the obvious reason that it would be very difficult to verify that oral communication occurred as intended. Since oral communication was the most important component of communication at Moura, this inability to audit it should have given some cause for concern. It did not appear to do so. The audit process had apparently not even picked up the fact that at weekends there was often no undermanager on duty and so there was no possibility of communication between undermanagers at change of shift. It is hard to escape the conclusion that the documentation was created primarily to satisfy the external assessors that the mine was deserving of QA accreditation and was never expected to have any real impact on the way things were done at Moura.

But the problem goes further than this. Neither the mining procedures document nor others like it had been seen by the undermanager-in-charge, even though he was responsible for underground procedures. He was asked about this at the inquiry and responded that it was not important for him to see these documents since they were designed to reflect what was already happening at the mine. In fact, however, they reflected what the writer *thought* was happening; as the inquiry revealed, the communication that was actually occurring fell far short of what was specified in these documents.

The apparent lack of concern by the undermanager-in-charge about the content of the procedures documents deserves further comment. Given that the instructions were that the documents be written so as to reflect practice, it is clear that he did not expect to have to adjust practice to bring it into line with the documented procedures. The logic is that any discrepancies between the document and the practice implied inappropriate documentation, not inappropriate practice; such discrepancies should be resolved by adjusting the policy, not the practice. It is noteworthy that this is precisely the advice given by Standards Australia in its handbook, *Quality Assurance Explained*:

> In rectifying differences between the quality manual/quality system procedures and actual practice, it is usually easier to rewrite than it is to change operating practices. (SA 1994:61)

Here then is an inherent limitation of QA. Under this system auditors are not expected to pass judgment on the procedures, merely to ascertain whether they are being adhered to. Thus, even with conscientious auditing, QA cannot be relied upon to assure safety.

Advocates of quality assurance will no doubt wish to object at this point that there is more to quality assurance than this. It is committed to the idea of continuous improvement (Voehl et al. 1994:14–16). When problems occur, managers must examine their system to see which procedures have given rise to these problems and amend them. Once these new procedures are in place and auditing shows that they are being followed in practice, quality will improve. Repeated again and again this process leads to continuous improvement. Books on QA are fond of illustrating this with a picture of a ball being pushed up an inclined slope. This is accompanied by an assertion such as: 'The prime task of any management team is to keep the total quality ball moving up the slope' (Voehl et al. 1994:16).

This elaboration of the QA system reveals that it is somewhat misnamed; it is not a system for guaranteeing quality but rather for guaranteeing the continuous improvement of quality. This is not just a linguistic quibble: when it comes to safety the distinction is vital. QA cannot assure us of safety; it can at best assure us that the management of safety is being continuously improved. This is entirely appropriate as a way of dealing with large numbers of relatively minor accidents. But it is not appropriate as a means of disaster prevention. Management cannot take the view that it will progressively improve its safety management system as disasters occur. Since disasters are so rare in any one organisation the opportunities for making improvements based on one's own experience are too limited to be made use of in this way. Moreover, even one disaster is one too many. Management must plan to avoid disasters altogether. They must get it right in the first place. The philosophy must be to learn from the experience of others, not simply from one's own. In short, the QA philosophy of codifying one's present practices and then seeking to continuously improve them is not an appropriate disaster management strategy.

In light of the preceding discussion it is not surprising that the QA system failed to prevent the Moura tragedy. Even if the system had been extended to cover spontaneous combustion there is no reason to think that the outcome would have been very different. The practices in relation to spontaneous combustion were far from satisfactory and documenting them for inclusion in the QA manual would not have changed this. In any case, the QA auditing process at Moura was simply a ritual that served to disguise problems rather than correct them. The Moura story is a stark illustration of the unreliability of QA as a tool for disaster prevention.

Lessons

What lessons can be drawn from the failure of all the audits at Moura to identify the problems that lead to disaster?

First, there must be appropriate systems to audit. As is by now very clear, high-risk organisations must carefully identify catastrophic hazards and painstakingly devise plans for the control of each hazard.

Second, good audits must evaluate the logic of these plans to ensure they are the best that can be devised. It is no good ensuring that procedures are being followed if they are inadequate procedures. One thing that any good system must be designed to do is note the early warning signs. Studies across a range of disasters show that there are always early warning signs, 'discrepant events' (Turner 1978:86), occurring in the lead-up to disaster, sometimes months in advance. These events develop and accumulate unnoticed or not understood. Disaster-prevention systems must be designed to highlight such events. A thorough audit of the system should identify what the early warning signs might be and then check that the system is designed to be sensitive to them.

Third, as described in Chapter 4, good systems have triggers—that is, events which, if they occur, result in some further predetermined action or investigation being undertaken. A good audit must find ways of pulling the triggers to see what happens. Thus, for example, if certain things are meant to happen when an alarm goes off, the audit process must set off the alarm and see whether they do happen. It is not enough to ask relevant personnel what happens when the alarm goes off, as happened with the gas monitoring audit. Again, if the system is designed to capture reports of smells it might be necessary to organise secretly for a deputy to report a strong tarry smell and see how the system responds.

A fourth principle that emerges from this discussion is the need to ask the following question. Suppose some aspect of the catastrophic risk management plan was not working—would this be uncovered in the audit? A creative auditor could imagine a number of critical ways in which the system might be failing. In the case of aircraft engine maintenance, suppose maintenance workers were not doing test runs. Would the audit uncover this? If the permit to work system is not well designed, would this be uncovered? If the plan specified that miners should be kept out of a mine when panels are going through the explosive range, but this was not the practice, would the audit reveal this? Of course, the possibilities are endless, and auditors could deal with only a few. But taking this approach would quickly generate a far more sceptical and thorough-going attitude than was revealed by any of the audits discussed in this chapter.

Finally, what is very clear from the Moura story is that there is a need to audit the auditors to ensure they are doing the job effectively. As Gunningham (1993) puts it provocatively in the title of an article on environmental auditing: 'Who audits the auditor?' The answer is that there must be a chain of auditing with the ultimate auditor being totally independent of the organisation and with a vested interest in ensuring quality auditing. The ultimate auditor might be a government agency, or a non-governmental public interest body—an environmentalist organisation in the case of environmental auditing, or perhaps a trade union in the case of health and safety auditing.

Government agencies are sometimes in a position themselves to carry out extensive audits of safety management systems, but in most contexts limited resources make this an impractical approach. Far better to concentrate resources on auditing the auditors and doing this in a sufficiently thorough way that the deficiencies of audits such as those carried out at Moura would have been laid bare. It is clear that, unless high-quality auditing is carried out regularly, even the best catastrophic hazard management plans run the risk of being no more than paper tigers.

7

WAS MOURA SAFETY-CONSCIOUS?

'Be fanatical about safety.' This was the instruction given to Moura mine management by a senior BHP executive some months before the explosion. It followed a double fatality at the adjacent Moura open-cut mine and mine management took this instruction seriously. The registered mine manager at Moura told the inquiry that he had got the message 'loud and clear' and when he became aware of a safety problem he dealt with it 'in a fanatical way'. He repeated, 'I considered myself . . . a safety fanatic.' A senior mines inspector testified that the BHP executive's words had 'projected downwards' a corporate commitment to safety which had previously been lacking at Moura. The result of this corporate commitment was that, as one mine deputy told the inquiry, 'everyone was always thinking safety'.

These were sincere statements; there is no doubt that management meant what it said. Indeed, in the next chapter I cite instances where production was sacrificed for the sake of safety. How then could Moura management be so neglectful when it came to spontaneous combustion? How could management believe itself to be fanatically safety-conscious and yet be so careless about catastrophic risk? This is the paradox that this chapter explores.

THE FOCUS ON LOST TIME INJURIES

To begin with we need to make a distinction. Explosions are relatively rare events—most mines have never experienced one—

but when they occur, they are likely to have severe consequences. These are low frequency/high severity incidents. On the other hand, slips, trips, falls, bangs and jolts are common events in mines, as they are in most industrial settings. Most mining injuries are caused in this way and many involve only a few days off work. These are high frequency/low severity matters.

Furthermore, careful records are kept of the number of lost time injuries (injuries that result in the loss of at least one day's work) and the lost time injury frequency rate (LTIFR) is often taken as a measure of safety performance. Concern about safety then becomes a concern about reducing the LTIFR. This is certainly what happened at Moura. By concentrating on high frequency/low severity problems the mine had managed to halve its lost time injury frequency rate in the four years preceding the explosion, from 153 in 1989/90 to 71 in 1993/94 (Windridge 1996: 13). By this criterion Moura was safer than many other Australian coal mines. But as a consequence of focusing on relatively minor matters, the need for vigilance in relation to catastrophic events was overlooked.

A similar situation was evident in the Westray mining disaster in Canada, described in Chapter 2. That mine had just won an award for reducing its lost time injury frequency rate when it blew up. Subsequent investigation revealed that the mine had been negligent with respect to catastrophic risk and had routinely violated safety regulations designed to prevent explosions (Glasbeek & Tucker 1993).

THE CORPORATE GOAL OF ZERO LTIFR

This focus on LTIFR was not just an idiosyncrasy of Moura mine management; it was BHP company policy. The long-term corporate target was a zero lost time injury rate, and mines were given twelve-month reduction targets. Most importantly, Moura mine managers understood the requirement to be fanatical about safety as a requirement to be fanatical about reducing the lost time injury rate. As the undermanager-in-charge explained, the earlier quoted senior executive

> was not satisfied with the performance of the operations across the board, all of the BHP Australia Coal operations, in regard to safety. The goal of our company was zero lost time injuries, and [in asking us to be fanatical] he was endeavouring to communicate to us on a

personal basis that he wanted us to do all in our power to achieve that.

Interestingly, this bias was built into the registered mine manager's remuneration system. His position description listed eight responsibilities, of which the first was

> to plan, organise and promote a safe working environment within the departments to achieve a no lost time injury frequency rate and communicate with all people to maintain safety awareness.

The fact that this was first in the list was intended to indicate the top priority of safety. Indeed, during performance appraisals for the purpose of salary review, this responsibility was rated priority A. All other responsibilities rated B, C or D.

But although this suggests that safety was a top priority, the explicit focus on achieving a zero LTIFR systematically directed concern away from issues such as spontaneous combustion. This is well illustrated by the registered mine manager's focus of attention after he returned from holidays, just a week before the explosion. He recounts how he was greeted at the mine gate by the safety board announcing that the mine had gone three weeks without a lost time accident. He was pleased with this achievement and it was one of the first topics he raised with the acting manager when he met with him for a briefing on events that had occurred while he was away. Several other topics were dealt with at this meeting but there was no mention of spontaneous combustion. The registered mine manager did not ask how the CO readings were progressing and was not told about the scare of 22 July at which a CO reading had indicated that a dangerous situation existed. (It will be recalled that this reading was subsequently dismissed as an error.) In short, the possibility that a heating was under way was not even raised. This is a telling illustration of the meaning of safety at Moura. In the following sections we explore in more detail some of the ways in which this focus on LTIFR distorted safety consciousness.

THE SYSTEMATIC RISK ANALYSIS

One of mine management's most significant initiatives, about three months prior to the explosion, was to carry out a systematic assessment of all the safety risks that might be encountered in the area to be mined.

Moura was one of the first mines to adopt formal risk assessment methods. In so doing it received high praise from the senior inspector of mines who described it as a 'quantum leap forward' in safety management. He wrote as follows:

> Management are to be commended for committing to these processes and in particular for embracing the greater involvement of the workforce. Those involved in the risk assessment did so in a very professional and objective manner and there were positive indications that the workforce as a whole is taking a greater interest (Inquiry exhibit 82).

The procedure involved a 'brainstorming session' in which a selected group of mine workers and officials was asked to list all the things that could possibly go wrong. Hundreds of possible risks were identified, of which spontaneous combustion was just one. The process involved giving each risk a numerical score based on two things: first, how likely the event was, and second, the potential severity of the event. Spontaneous combustion was judged to be relatively unlikely and furthermore *unable to cause harm to workers*. It was thus scored as a relatively low priority hazard.

The big fallacy in this process is obvious: while spontaneous combustion, by itself, might not cause harm, if there were to be a concurrent build-up of methane gas it could trigger an explosion with the possible loss of many lives. This possibility was simply not encompassed by the risk analysis undertaken.

Not only was the group asked to identify and assess risk; it was also asked to identify how risks could be controlled. The controls identified in the case of spontaneous combustion were, first, that the panel life was short and, second, that continuous gas monitoring was in place. A possible extra control was that water could be pumped into an area of suspected heating. These ideas were simply thrown up in the process of brainstorming, but not seriously evaluated. As we now know, the short life of the panel is no guarantee against spontaneous combustion and the fact that continuous gas monitoring is being undertaken is no guarantee that the results will be understood or responded to appropriately. Nor can it be assumed without careful consideration of the mine layout that flooding with water is feasible. Given the drought, there may not have been enough water available, anyway. Thus, the end result of the risk analysis was to generate a false sense of security that everything was under control with respect to spontaneous combustion.

Clearly, attention was spread too thin. Rather than trying to think carefully about a small number of catastrophic hazards, the net was cast in such a way as to cover every conceivable risk of injury. This was an analysis that aimed for breadth rather than depth, but in aiming to be comprehensive it was necessarily superficial. This was the inevitable outcome of a focus on lost time injury.

SAFE BEHAVIOUR INVOLVEMENT

Another management tool used at Moura to improve safety was the safe behaviour involvement program (see Whiting 1993). The aim is to identify frequently occurring behaviour that creates a risk, to ask that the workers change this behaviour, and then to monitor the changes. The simple fact of monitoring and publicising the frequency with which problematic behaviour is occurring leads, over time, to improvements, since workers become involved in monitoring their own progress.

Here is an example of how this strategy was used at Moura. High-pressure water hoses that are snap-locked together can be extremely hazardous if they become uncoupled and whip around. There had been several reports of this occurring at other mines. To combat this danger, good practice requires safety clips to be fitted. An initial survey revealed that only 66 per cent of couplings were safety-clipped in this way. An educational campaign was mounted and over the next few weeks the compliance rate was monitored and rose progressively until it was over 90 per cent.

Safe behaviour programs of this nature seem well suited to the task of reducing numbers of routine injuries. They are less easily adapted to deal with rare but potentially catastrophic events. Where the behaviour in question is associated with the rare event, such as behaviour in response to an alarm, it cannot be monitored on a regular basis. The challenge is to identify disaster-preventing behaviours that are supposed to occur routinely and that consequently can be monitored regularly. This is not impossible. But it is easier to identify unsafe behaviours with the potential to cause more ordinary kinds of injury. And given Moura's focus on reducing lost time injury, that is how this tool tended to be used. Certainly, there was no attempt to use safe behaviour involvement to deal with the risks of spontaneous combustion.

A second example of the use of safe behaviour involvement reinforces this point. It concerned the eradication of smoking in the lamp room. This was an above-ground room that housed the miners' safety lamps; it also functioned as an unofficial meal room. Miners often smoked during meals, despite a No Smoking sign. Initial observations indicated that compliance with the no smoking rule was about 60 per cent. A monitoring program began and over time 100 per cent compliance was achieved. It should be noted, however, that while smoking underground in a gassy mine such as Moura could potentially ignite an explosion, smoking in the above-ground lamp room involved no such risk. This was a safety campaign that had no direct connection with the control of catastrophic risk. It could be argued that the anti-smoking campaign promoted safety awareness generally, but while this may be true it did nothing to focus attention on catastrophic risks.

Safety meetings

A further aspect of the commitment to safety at Moura was the network of safety meetings and committees. A joint worker–management safety committee met regularly. It was, however, largely reactive, in the sense that it responded to accidents that occurred and explored ways of preventing their recurrence. The committee was particularly concerned about manual handling injuries caused by the operation of roof-bolting machines and was seeking ways to reduce the amount of handling involved. Its stated ultimate purpose was to achieve zero lost time injuries. This focus on LTIs worked against any consideration of disaster prevention. Thus, for instance, the crucial safety question of whether men should go underground after a panel was sealed was never raised on this committee!

Management also organised regular mass safety meetings at which the latest LTI trends were presented, recent accidents or incidents analysed and other safety information communicated to the men. The men were also invited to raise issues from the floor, which they invariably did. These were always given serious consideration by management. However, the possibility of an explosion ignited by spontaneous combustion did not come up at these meetings.

One matter with the potential to cause disaster *was* on the minds of management and *was* discussed at these meetings. This was the frequency with which electricity cables in the mine were being accidentally cut by moving vehicles, causing 'flashes' capable of igniting any methane that might be present. The minutes of a meeting two months before the explosion report in capital letters:

> OUR PERFORMANCE IN RESPECT TO THE NUMBER OF CABLE FLASHES IS FAR FROM SATISFACTORY. ALL EMPLOYEES ARE URGED TO EXERCISE CONSTANT VIGILANCE IN RELATION TO THEIR ACTIONS, OBSERVATIONS AND AWARENESS.

This concern stemmed in part from the fact that flashes were so obviously dangerous, but also from the fact that the inspectorate was highly critical of what was happening and was demanding improvement.

Interestingly, despite the above plea to employees to be more careful, the issue of cable flashes demonstrates the limits of management's commitment to safety. Underground shuttle cars are powered by electricity supplied by trailing cables. A reel on the back of the car is supposed to pay out or reel in cable as the car moves. If it fails to do so, in particular if it fails to reel in cable as the car moves backwards, the shuttle car is likely to run over the cable, wrenching or cutting it and causing a flash. The reel can fail to move if, for instance, it gets clogged with mud. In the seven months prior to the explosion there had been six cable flashes, most of them caused by shuttle cars running over their trailing cables. It is clear that urging miners to be more careful, as the above statement does, is not the solution. This is yet another example of the widespread tendency to hold workers responsible when things go wrong, a singularly unhelpful approach (for an extended discussion, see Hopkins 1995, Chap. 1).

In this case a far more effective strategy was at hand: fitting the shuttle car with an alarm that is activated when the vehicle moves but the reel fails to move. Only about 50 per cent of shuttle cars had been fitted with these alarms. If safety really was the top priority it was said to be, one might have expected that shuttle cars without this device would be taken out of service and appropriate devices fitted as a matter of urgency. Here was a case, then, where management was certainly not behaving in a fanatical way about safety. Perhaps it is precisely because these flashes were not generating lost time injuries that management was as relaxed about them as it was.

TRAINING

Some concern for safety was also evident in the mine's commitment to training. But training was primarily about the safe operation of mobile machinery—particularly diesel vehicles. In addition, the safety/training manager was responsible for refresher training on a variety of topics. Emergency evacuation training was carried out regularly and was credited with saving the lives of the ten men who actually survived the Moura explosion. However, the safety/training manager did not view spontaneous combustion as a major risk at the mine and accordingly had not organised a refresher course on this subject. It had, he said, a low priority on his list of things to do.

THE FATE OF ONE SPONTANEOUS COMBUSTION SAFETY INITIATIVE

Earlier chapters have demonstrated just how inattentive management was when it came to spontaneous combustion. It was perhaps predictable therefore that an initiative by one man who *did* take the issue of spontaneous combustion seriously would come to nothing. The details are worth recounting.

One critical indicator of the possibility that a heating is present is the rate of CO production in litres per minute. The remote gas monitoring system measured CO in parts per million. To calculate the rate in litres per minute one needed to know also the velocity of the airflow at the point of measurement. Had the remote system been able to measure this automatically it could have produced the required figures automatically. But it didn't. Accordingly, deputies had to go down to the monitoring points and measure the air velocity with hand-held instruments. Once this information was available the necessary figures could be calculated manually. Reliance on manual calculations proved to be one of the factors that contributed to the disaster, for these calculations were simply not done and the crucial data were therefore not available.

In 1992 the then registered mine manager, a man who was more aware of the issue of spontaneous combustion than most, requested his superiors to authorise funds to install permanent velocity meters that would feed their data automatically to the surface computer. His superiors apparently failed to act on this request and the initiative had not been implemented when he left his position shortly afterwards. Nor was it implemented in succeeding years. Here again management, as a whole, was anything but fanatical about safety when it came to questions of spontaneous combustion.

THE DANGER OF FOCUSING ON LTIFR: FURTHER REFLECTIONS

The Moura (and Westray) experiences cast doubt on the often repeated claim that a good LTIFR indicates that safety is being well managed or, alternatively, is an indication of a high level of safety awareness among the workforce (this claim was made at Moura). Let us examine why, a little more broadly.

It is widely recognised that where organisations are judged by their LTIFR they will resort to ways of reducing this figure that have nothing to do with improving safety. More effective claims/injury management, for instance, will result in workers who previously took several days off work being treated and brought back to work on the next shift, perhaps on alternative or light duties. The result is that they no longer count as lost time injuries. Effective claims/injury management, in other words, results in the conversion of lost time injuries to non lost time injuries. The statistics can thus improve with no reduction in the real rate of injury. This is all quite legitimate. But what it means is that the LTIFR becomes a measure of how well injuries are being managed, not how safely the organisation is performing. (For an extensive discussion of this phenomenon, see Hopkins 1995, Chap. 10.)

There are also several less legitimate strategies for reducing the LTIFR, ranging from bringing people back to work who, on medical grounds need to be at home—the so-called walking wounded—through to threatening to sack workers who suffer injuries or make compensation claims (see Hopkins 1995, especially Chaps 3, 9).

The chief inspector of coal mines spoke eloquently about this at the annual colliery safety symposium in New South Wales in 1993:

> It is curious that in the last ten years the LTIFR for the industry has been reduced to approximately one third of its former value, yet the number of reported serious bodily injuries, normalised to

> industry employment, has barely reduced, if at all. One can only conclude that these two indices are a measure of fundamentally different things.
>
> There can be no doubt that the LTIFR is not just a measure of how safety has been, and is being, managed. It is also a measure of the industry environment, and of the workers compensation environment. Unfortunately, in addition to reflecting managing safety the LTIFR also reflects efforts to simply manage the measure, with no regard to safety.

The preceding discussion shows how claims and injury management can reduce the LTIFR substantially without any corresponding improvement in safety. But beyond a certain point further progress depends on attending to safety itself. At this point, further reduction in the LTIFR will indeed be an indication of safety improvements.

But here is where the main problem identified in this chapter comes into focus. It is painfully obvious that a good LTIFR is at best an indication that the processes leading to lost time injuries are being well managed, not that low frequency/high severity risks are being carefully controlled. On the contrary, the danger is that a single-minded focus on reducing the LTIFR leads systematically to the neglect of catastrophic risk. Where there is potential for catastrophe, safety management must not be driven exclusively by a concern to reduce the lost time injury frequency rate. That way lies disaster, quite literally.

Conclusion: the need for appropriate incentives

This chapter began with a paradox: how could mine managers claim to be fanatical about safety and yet be so careless of the possibility that a spontaneous combustion might trigger an explosion? The explanation of this paradox is that mine management was focused on reducing the lost time injury frequency rate, which systematically distracted attention from catastrophic risk. It is not that managers were careless about safety; they were, on the contrary, actively involved in making the mine a safer place. But the mine's incentive structure was one that focused on lost time injuries and provided no rewards for the control of low frequency/high severity risks.

Here then is perhaps the most vital lesson to be learnt. Corporate headquarters must find ways to reward managers who attend conscientiously to the control of catastrophic risks. Good performance in this respect is of course much harder to measure than success in reducing lost time injuries. But unless this lesson is taken to heart nothing is more certain than that, over time, management will again become complacent about catastrophic risks, as it has so often in the past.

8

Production Before Safety?

At the very moment the mine exploded, the shift undermanager was at the surface, on the phone to a production crew underground. The line went dead. This was the last contact anyone had with that crew; they were never seen again. Inevitably, then, those last words take on a special significance. What were they about?

It will be remembered that the question of whether it was safe to send the men underground that night had been raised and resolved affirmatively as far as management was concerned. Nevertheless, the mine was about to enter the explosive range. Were there any lingering doubts about safety in the shift undermanager's mind? Was he on the phone about safety? On the contrary. His concern at the moment of the explosion was production. Coal produced in the mine was sent to the surface on a conveyor belt and the tonnage coming out of the mine was carefully recorded on a tonnage meter in the manager's office. The shift undermanager had noticed that production was down and had rung to discover why. There had been a problem with the mining machine. This was the topic of conversation when the line went dead.

By itself this can hardly be taken as evidence of a tendency to give production priority over safety. But is it in some way symptomatic of such a tendency? This is a question that needs careful consideration; that is the purpose of this chapter.

Production pressures

The first point to establish is that there was indeed a pervasive pressure to produce as rapidly as possible. The 512 panel had been designed to achieve—and did in fact achieve—a rate of production higher than any previous panel at Moura (Windridge 1996:14). This was a matter of which management was intensely proud.

Daily production targets were set and if they were not met management went to considerable pains to find out why not and what could be done to increase the rate of production. Shift undermanagers had to report the tonnage produced at the end of every shift and to note the amount of down-time and the reason for it.

Perhaps most dramatically, the undermanager-in-charge attended a meeting with the men a few months before the explosion where the future of the mine was called into question: the men were told that the mine might be forced to close unless productivity increased. This situation had developed because coal prices had dropped substantially. There could be no greater production pressure than the threat of mine closure and consequent job loss.

The men themselves took a keen interest in the mine's output. In part, this was because they were paid a bonus that depended on how much the mine produced. The amount of the bonus was very substantial and in good times could double a miner's take-home pay. One miner told the inquiry that he used to go to the manager's office at the end of every shift to check the tonnage for the shift so that he could get some idea of how much bonus had been earned. Production crews even competed with each other to see who could cut the most coal.

It should be noted that managers received the same production bonus as the miners. This would have added to their incentive to keep the production rate as high as possible.

Safety before production

Despite these pressures it is not hard to find instances where safety has taken precedence over production. The inquiry was told of two occasions on which deputies had stopped production because methane gas readings were too high. On one occasion the production crew was able to solve the problem and return to work, while on the other the crew had to withdraw altogether and wait for the shift undermanager to come underground and organise ventilation changes to improve the airflow.

One of the most interesting examples of safety taking priority over production was the issue of taking 'easy coal'. The coal seam being mined was 4.5 metres thick. In previous panels the full thickness of the seam had been mined which meant that miners were exposed to high walls of coal as they worked. Falls of coal from these walls were not uncommon with the result that one man had recently been killed and, on a separate occasion, another had suffered a badly broken leg. Because of this, a decision had been made not to mine the seam to the full depth. In the first or development phase of mining in the panel, only the top 3 metres of the seam would be mined. In the second or extraction phase, the lower 1.5 metres of the seam would be partially extracted in such a way that miners were not exposed to a wall of height greater than 3 metres. Of necessity this meant that a considerable amount of 'easy coal', coal that might have been extracted with minimal effort, would be left in the lower 1.5 metres of the seam. This was a management decision that very definitely sacrificed output for safety.

Miners did not always observe this restriction. From time to time they succumbed to the lure of easy coal and in so doing exposed themselves to a greater risk of injury. The reason they did this was quite simply that taking easy coal increased the mine's output and hence their bonus. This was acknowledged by more than one inquiry witness. On one occasion the requirement was extensively violated over two shifts and was discovered by the registered mine manager. He called in the production crews for all shifts and warned them about the practice, saying that if they did not comply with the extraction plan they would be stopped from taking any coal from the bottom 1.5 metres of the seam. This would certainly have impacted on the production rate and on the men's bonuses. It was widely believed that the registered mine manager meant what he said and the men thereafter complied with the requirement.

The issue of cable flashes was portrayed at the inquiry as another matter in which safety took priority over production. A cable flash, it will be remembered, is sparking caused by shuttle cars running over and damaging their trailing electric cables. There had been 22 cases of cable damage (not all flashes) in the first six months of 1994. The registered mine manager had developed a policy requiring that, when a flash occurred, mining would stop until the matter had been fully investigated by mine management and the inspectorate. This might mean that production was stopped for a shift or, on occasion, for two or three shifts.

But this was not a straightforward case of safety having priority over production. Because they are so dangerous, cable flashes must be notified to the inspectorate. The inspectorate was very concerned about the situation and was pressing management to take urgent action. The previous practice when a cable flash occurred was to replace the cable and continue mining. The new policy, involving the cessation of mining until an investigation had been completed, was a result of pressure from the inspectorate. It appears that, left to itself, management had not accorded the issue of cable flashes top priority. As we concluded in the last chapter, this was a matter in which management was somewhat less than fanatical about safety.

Production before safety

There were undoubtedly situations where safety had priority, but there were numerous ways in which the reverse was true. These were often quite subtle and unrecognised. We have already seen one example of this in Chapter 3. Safety reports were prepared at the end of every shift by the most junior members of the management team, the deputies; production reports were prepared by their superiors, the shift managers. The very structure of this reporting system gave production precedence over safety.

The bonus system was another area where this priority was built in. It is true that bonuses were paid to miners for safety performance. But these were relatively small and paid at relatively infrequent intervals. In contrast, the production bonus was paid on a weekly basis and was a substantial part of the miners' income. Here was a built-in incentive to give production precedence. It is hard to imagine how this could have been otherwise. Safety bonuses are usually paid on the basis of the number of lost time injuries. One of their effects therefore is to discourage the reporting of injuries; increasing the size of such bonuses would increase the incentive not to report. Large safety bonuses paid on the basis of LTI figures are therefore not to be recommended. On the other hand, large production bonuses are a fundamental feature of the coal industry. In short, the system of financial rewards gave top priority to production in an inescapable way.

Another subtle indicator of priorities is the status of safety meetings. At least one shift undermanager took the view that these meetings were the way you filled in time when mechanical breakdown or some other reason brought production temporarily to a halt. He is reported to have said that he hoped the men would go underground as usual while the mine was explosive, otherwise they would have to 'do safety meetings for nine hours'. The use of safety meetings to fill non-production time is perhaps not a matter of substance but it is difficult not to read it as a symbolic statement of priorities.

Perhaps the best example of the priority of production was the decision to send the men underground on the night the panel atmosphere was explosive. Let us consider this issue in more detail. It takes two things to cause an explosion: an explosive gas mixture and an ignition source such as a spark or spontaneously burning coal. Explosions can thus be avoided by ensuring either that there is no explosive gas mixture present, or that there are no ignition sources. A prudent policy seeks to address both factors simultaneously. Coal mine regulations generally embody this prudent policy, seeking simultaneously to control both ignition sources and explosive gas build-ups. In a sense this involves two lines of defence; if one fails the second line should hold, all going well.

Prudent policy seems not to apply in the case of panel sealings. Although it is known that the panel will at some stage contain an explosive mixture of gases, it is presumed that there will be no ignition source present and that it is safe to send men underground to work in other parts of the mine. At such a time there is only one line of defence. Should that defence fail, should an ignition source in fact be present, an explosion will occur. That is what happened at Moura.

The safest policy would be to keep the men out of the mine while a panel was going through the explosive range. This may mean sacrificing one or two days of production. In this matter, Moura management opted for second best in terms of safety in order to maximise production. Here was a clear and ultimately disastrous case of putting production ahead of safety.

Consider, finally, the culture of denial described in Chapter 4. There was at Moura an extensive set of beliefs which denied that there could be any significant risk of spontaneous combustion and which nullified whatever evidence there was to the contrary. So powerful was this set of beliefs that it amounted to a culture of denial. The psychological origins of this culture are clear. Had the indicators of spontaneous combustion been taken seriously and acted on, this would have entailed the disruption of production and quite probably the closure of the panel for an extended period of time. Given that this was such a high-production panel and that the mine was operating under intense pressure to produce, this was to be avoided at almost any cost. The culture of denial was a mechanism that allowed safety concerns to be discounted and production to take precedence. This is not to say that it was invented for this purpose; this was not a deliberate or intentional process. It is simply that these beliefs were accepted uncritically because of the functions they performed in allowing production to continue.

DID MANAGEMENT CONSCIOUSLY PUT PRODUCTION AHEAD OF SAFETY?

The question that remains unanswered is whether or not management consciously put production ahead of safety. Did they recognise what they were doing? Counsel for the miners apparently thought they did. He submitted to the inquiry

> that BHP ran a dangerous race against time to extract all of the coal from 512 [panel] in accordance with the plan. It lost that race, eleven men were killed as a result.

But the evidence suggests that this is too harsh a judgment. The culture of denial was a mechanism that shielded all concerned from recognising the priorities they were, in practice, implementing. As for the decision to send men underground while the panel was explosive, this was common practice, and while it was clear to all that there was a production imperative involved, there was no clear recognition that this policy fell considerably short of best safety practice. The notion that at least two lines of defence are necessary to safeguard against explosions—a notion that was accepted in routine mining situations—seemed not to be part of management consciousness when it came to panel sealings. It is not that management perceived a conflict between production and safety and opted for production; they simply did not perceive themselves as having to make that choice.

On the other hand, where a conflict was perceived, it was safety that took precedence. That at least is what happened in relation to the mining of easy coal. Here was a matter where management was acutely aware of the possibility of injury caused by falls of coal from high walls. This was a lost time injury matter and, as such, as we saw in Chapter 7, of great concern to management. In this case, then, production was consciously sacrificed for the sake of safety.

Let us return to the incident at the beginning of this chapter, the telephone call to the production crew at the time of the explosion. It is now clear that the incident was highly symptomatic. Production was indeed closely monitored and crews were under pressure to produce as much as possible, as the phone call implied. These were matters to which great attention was paid. On the other hand, the risk of spontaneous combustion was not effectively or closely monitored and very little attention was given to the wisdom of sending men underground that night. It seems undeniable that production had a higher priority than safety on the night in question. This is not to say that the shift undermanager was conscious of these priorities. He seemed genuinely unaware that safety had been sacrificed to production. But the fact that, at the moment of the explosion, he was on the phone to the men who died, asking about a production hold-up, is indeed an eloquent statement of the priorities at Moura.

9

THE ROLE OF BHP

'We accept accountability for the safety of our people and accept accountability for the disaster.' So said a senior BHP executive in the wake of the Moura disaster (*Australian*, 16 January 1996). Just what this statement meant, however, is difficult to determine. Certainly BHP has paid compensation to relatives. But neither BHP nor any of its managers or directors has been prosecuted for what happened. Thus senior management, and in particular the chief executive officer, has not been held accountable at law.

All this raises the question of just what is BHP's role and responsibility for disaster prevention. To answer this we need first to know something about BHP.

At the peak of the BHP pyramid is The Broken Hill Proprietary Company Ltd, listed on the stock exchange and owned by tens of thousands of shareholders. This is a company with worldwide interests but with its head office in the BHP tower in Melbourne. BHP Ltd owns a number of subsidiary companies, among them BHP Steel, BHP Petroleum and BHP Minerals Pty Ltd. BHP Minerals, in turn, owns a number of subsidiary companies, among them BHP Australia Coal Pty Ltd, with headquarters in Brisbane. This is the company that, at the time of the explosion, operated the Moura mine, as well as a number of other coal mines in Queensland. The company now operates BHP's coal mines in New South Wales as well. It has recently added mining operations in Indonesia to its portfolio and dropped the word Australia from its name. However, it will be referred to here by its name at the time of the explosion.

Decentralising responsibility for safety

How is safety managed in this corporate structure? The policy is that, as far as possible, responsibility for safety is decentralised to the plant or mine site. The result is that safety has a remarkably small organisational presence in this corporate structure. At the Melbourne headquarters of BHP Ltd there is one safety specialist with perhaps two support staff. Some years ago this safety unit had a staff of thirteen but this was cut right back as a result of a decision to devolve responsibility for safety.

The role of the corporate OHS manager is to advise the BHP Board, among other things, on whether safety audits are being done within the company and whether the information coming from these audits is reliable. This enables directors to satisfy the legal requirement that they exercise 'due diligence' in respect of safety. The legal advice on which the Board operates is that, if the OHS manager can report to them at each meeting that safety systems are in place and being properly audited, this discharges their personal obligation in relation to safety.

The view from BHP headquarters is that it cannot be prescriptive about safety; that is, it cannot tell the many site managers in its far flung empire what to do. The role of head office is to encourage local managements to have safety systems but not to prescribe in detail what those systems should be. Likewise, head office requires managers to set themselves performance goals in relation to safety, without specifying what those goals should be. (As it turns out, these goals are usually set in terms of reductions in lost time injury rates, which, as previously argued, systematically distorts the safety effort.) BHP head office does have a safety policy but, in line with the general philosophy of devolution, this is a one-page statement of commitment to safety which of necessity is extremely general. This statement was issued by the chief executive officer six months before the Moura disaster.

BHP Minerals is one step down from BHP Ltd. It, too, is headquartered in the BHP tower in Melbourne and is an umbrella entity to which a number of BHP resource companies around the world belong. Like its parent company, BHP Minerals has a minimal role with respect to safety. It has two or three safety specialists and two or three support staff, divided between its Melbourne and San Francisco offices.

A further step down the chain is BHP Australia Coal (BHPAC) with headquarters in the Riverside Centre tower in Brisbane. It has a safety team of about four people, including secretarial support. The coordinator of this unit is answerable to the human resources manager who in turn reports to the general manager. The organisational profile of this safety unit is thus quite modest. As in the case of BHP and BHP Minerals, its role is one of general policy development and the encouragement of safety auditing at the various company sites. Thus, in March 1994, five months before the explosion, BHPAC put out a set of sixteen safety policy standards, each about a page in length, covering such matters as accident investigation, safe working procedures, auditing and so on. Spontaneous combustion was not addressed in these standards, nor were issues of gas management or roof fall. The safety unit also produces statistics comparing the LTIFRs at the company sites.

Prior to the explosion, BHPAC had instructed all its sites to develop safety management plans. Moura was one such site and the plan covered issues generic to both the open-cut and underground operations at the site. It thus said nothing specifically about spontaneous combustion, methane gas control or roof fall—that is, it made no mention of catastrophic risks.

Six months before the Moura explosion the BHPAC safety unit had conducted a so-called executive or corporate safety audit of its seven Queensland sites. Interestingly, despite the general view that head office should not be prescriptive, the report from this audit did include a few detailed prescriptive statements—for example: 'It is requested that mines require the wearing of life jackets when employees are working around a half a metre or more of water including in row boats and on pontoons.' But on the whole the report reflected the general philosophy that the role of head office was to advise rather than require and to deal in generalities rather than specifics. Thus the report set up a task force to make recommendations on lock-out procedures (procedures to isolate pumps, taps, valves etc on which maintenance work is being done, so as to prevent their accidental use). But it noted that 'the ultimate decision (on implementing these recommendations) will lie with mine/port managers as a group'.

Again it noted that, at Moura, 'improvements in both capital equipment and procedures must be achieved . . . Also, proper support systems for extraction need further assessment'. The audit report said nothing about spontaneous combustion or gas management, indicating yet again a systematic neglect of catastrophic mine explosions. It should be noted finally that the above statements reflect the view that, on the whole, responsibility for the details of safety management remains with the site managers.

The philosophy of devolving responsibility to site level, coupled with minimal size of safety advisory groups at higher levels, meant that BHP head offices were not perceived on site as playing any role in relation to safety. Consider the comments by Moura's safety and training manager during cross-examination at the inquiry:

> What I'm trying to get at is if an organisation as big as BHP doesn't have some central advice to individual mines with regard to training of personnel?
>
> *I'm sorry, I can't answer that question. I do not know.*
>
> So what are you saying? It was left largely to the individual mine, in your case Moura, to design and implement your own schemes?
>
> *Yes, we were based on statute. We followed statute and on a needs basis in line with the corporate company policies.*
>
> In your time at Moura, apart from the audit that you've referred to, did you ever get a visit from people who were in the safety department of BHP or any directive or advice from headquarters relating to mines safety?
>
> *I cannot recall any, no.*

If Moura's safety and training manager was unaware of directives or advice from any BHP safety department, clearly little or no advice or instruction from that source was arriving at the site level. Picking up on this silence, a government inspector commented at the inquiry on BHP's relative lack of 'corporate commitment to make things happen' in the safety area in the early 1990s.

Consistent with this policy of decentralisation, the general manager for the combined Moura open-cut and underground operation held the view that he had no particular responsibility for the safety of the underground pit (see Appendix 1 for management hierarchy). He admitted at the inquiry that he had no expertise on the question of spontaneous combustion. He did occasional inspections in the underground mine but these were not safety inspections; 'just a case of having a look round at the way the operation was running, introducing myself to the miners and officials, discussing with them what they thought of the operations and the prospects for the underground mine'. No mention of safety here; only an implicit threat that if the mine did not maximise production, its future might be in doubt.

In summary, BHP's policy of decentralising responsibility for safety meant that as an organisation it had very little safety expertise, was not in a position to provide detailed advice on safety issues at its various sites and was perceived as playing no particular role in site safety, other than from time to time urging sites to improve their safety performance and providing some limited incentives for them to do so. The dreadful irony of all this is that, as we saw in Chapter 5, management of the underground mine continued this policy of shifting responsibility for safety down the hierarchy to the point where, on the night of the explosion, it seemed to end up as a policy of every man for himself.

The inappropriateness of decentralisation of responsibility for disaster

In reflecting on the appropriateness of this decentralisation of responsibility for safety we need to note first that it is not only in this area that a policy of decentralisation prevails. The policy also applies to the company's core goal, profit making. BHP Ltd does not tell its subordinate companies in any detail how to maximise their production or their profits; merely that they must do so. If they fail to perform, there are consequences, ranging from loss of bonuses or promotion prospects to site closure or sell-off. Moreover, the data on production and profit are available on a regular basis to head offices who are thus able to monitor closely the extent to which the component parts of the BHP empire are performing as required. Decentralisation in this matter therefore involves no reduction in the level of scrutiny or the level of pressure that head office can bring to bear on poor performance. In short, it involves no reduction in the extent to which site managers are held accountable.

The situation is similar with respect to safety as measured by lost time injury frequency rates. These data can be made available on a regular basis to head offices and the performance of individual sites can be closely monitored. In this way BHP is able to exert effective pressure on its site managers to improve their LTIFR performance.

None of this is true for the management of catastrophic risk. Most of BHP's sites have never suffered a disaster and many years may go by without a disaster anywhere in the BHP empire. It is therefore impossible to compute a disaster frequency rate for an individual site in the way that an LTIFR can be computed. There is thus no simple or direct way in which head office can monitor the success with which site managers are controlling catastrophic risk.

It might be objected that, where sites have catastrophic risk management plans in place, then the auditing of these plans provides one way to monitor how well risks are being controlled. But as was demonstrated in some detail in Chapter 6, auditing is notoriously unreliable. Previous disasters or potentially disastrous occurrences have frequently been preceded by audits that should have revealed serious inadequacies, but failed to do so.

In this context, then, the devolution of responsibility means that head office no longer has a reliable way of knowing whether this responsibility is being exercised conscientiously. In short, it has no way of holding site managers accountable in this respect.

It is axiomatic that managers who are held accountable for production and for lost time injury rates, but not for the control of catastrophic risk, will tend to overlook the latter. Here then is one good reason why responsibility for the management of catastrophic risk should not be left to site managers but should be effectively exercised from head office. The logic of this argument is that head offices should maintain large safety departments, whose prime concern is the management of catastrophic risk and who are in a position to give detailed technical advice on such matters, to issue instructions or advise chief executive officers on instructions to be issued, and to scrutinise closely the level of compliance with these instructions.

Further arguments for centralised control of catastrophic risk

There is a second and related reason why it is appropriate for head office to assume control in this matter. Because most mines have never had an explosion, individual mine managers have no reservoir of experience to draw on—no direct history to serve as a warning. The same is not true for BHP as an organisation. According to counsel at the inquiry BHP has had five mine explosions or major fires—three in New South Wales and two in Queensland. Prior to Moura 1994, the most recent of these was the explosion at the adjacent Moura No. 4 mine in 1986 in which twelve men were killed. Preceding that was the explosion at Appin in New South Wales in 1979, in which fourteen miners died. BHP as an organisation is in a position to take note of this history and to learn from it in a way that individual mines are not. It is thus in a good position to exercise leadership, in fact control, and to instruct mine managers in clear and detailed terms about what they need to do to avoid such disasters.

An example of the kind of control that head office might exercise would be to issue instructions about the circumstances in which workers should be withdrawn from mines. Such an instruction might well specify that miners are not to be underground when a panel is going through its explosive range. It is noteworthy that the chief inspector of coal mines in Queensland recommended at the inquiry that BHP take control in this matter of when persons should be withdrawn, rather than leaving it to mine management with the risk that they will adopt some 'feel good or . . . vague position'.

A further argument for centralised control concerns efficiency. As was amply demonstrated by the Moura events, the effective control of a hazard like spontaneous combustion requires the attention of someone who is expert in this area. Rather than have one such expert employed at each mine site, it would make good sense for such a person to be employed at head office and to exercise control in relation to this hazard at all underground coal mines in the group. This kind of specialisation would enable one person to embody the wisdom of the organisation on this issue. Such an expert could visit each underground mine on a regular basis to ensure that the risk of spontaneous combustion was being effectively managed. Should capital expenditure be necessary, this person would be in a position to make recommendations to head office for special appropriations for this purpose. One such expert might be appointed at head office for each of a small number of recognised catastrophic hazards.

Where in the hierarchy should expertise be located?

If these arguments are accepted the question then arises as to where in the corporate hierarchy this safety expertise should be concentrated. BHP Australia Coal is an obvious site since it is in a position to deal with disaster prevention in Australian coal mines generally. But BHP Minerals is another candidate since it is responsible for all BHP coal mines, both in Australia and overseas.

The purview of BHP Ltd is even wider. Consider a problem that arose in the BHP petroleum division in relation to offshore production involving tankers (Barrell 1996b). As tankers are emptied of petroleum the empty space may fill with inflammable fumes. If the gas mixture contains a certain amount of oxygen the mixture is explosive. Should there be any static electricity generated in the operation there is a serious possibility of explosion. Safety precautions are aimed both at minimising the risk of spark, by proper earthing, and at ensuring that the gas mixture never enters the explosive range, by pumping an inert gas into the hold. On two occasions, in 1995 and 1996, there were conscious violations or potential violations of these safety precautions. Fortunately, no explosion occurred. But the matters were so serious that several inquiries were conducted, both by BHP Petroleum and by regulatory agencies. The issue was even raised in the Senate.

These matters seem eerily similar to events at Moura, yet they occurred in a part of the BHP empire organisationally and geographically quite remote from Moura. Only the view from head office in Melbourne would be sufficiently broad to recognise the similarities and see that there might be advantages to a corporate-wide approach to disaster prevention. Perhaps there is an argument for different kinds of expertise to be concentrated at different points. But wherever located, it seems clear that there is a need for concentrations of expertise and control above the mine or plant level.

Let us note in passing that this argument about the need for corporate control in relation to disaster prevention is also an argument for strong government intervention in relation to disaster control. The Queensland government has presided over four coal-mining disasters in 22 years with a total of 53 lives lost. Obviously, governments are in a position to draw the lessons from this experience in a way that individual mines cannot. This is a strong argument for governments to exercise detailed and quite prescriptive controls when it comes to disaster prevention. This point is developed further in Chapter 11.

Developments since Moura

BHP has taken numerous steps to improve safety since the Moura tragedy. One is the adoption of the NOSA system of safety management throughout much of BHP Minerals. This is a general system, not focused on catastrophic risk. Within BHPAC each mine now has a safety department of up to five people implementing the NOSA plan. From the perspective of this chapter, this represents only a very tentative first step by BHP Ltd in the direction of exercising greater corporate control.

Beyond this, BHPAC has developed a catastrophic risk management strategy, requiring each site to identify a small number of catastrophic risks and develop plans to control them. It has developed detailed guidelines on how to do this but it is still up to individual mines to develop their own plans. These plans are now subject to intensive auditing by external auditors from other parts of the corporate empire. Again, this is a step in the direction of more centralised control of catastrophic risk, but it is only a step. It is clear that mines retain ultimate control and ultimate responsibility.

It is worth illustrating this point. One of these intensive audits revealed that the mining machines at a particular mine were not equipped with methane gas monitors. The mine's defence was that this was not a gassy mine and hence the monitors were unnecessary. The counter-argument is that, as mining progresses into new areas, methane gas might be encountered unexpectedly. It would thus be prudent to fit gas monitors. The mine concerned rejected the recommendation of the audit report that it fit monitors. A more centralised approach might have seen the mine instructed to fit the necessary items, perhaps with the costs being met from a central budget.

If the arguments of this chapter are accepted, it is not enough to leave mine managements with ultimate responsibility for disaster prevention. If further Mouras are to be prevented, the BHP corporate hierarchy needs to take responsibility in a way that it currently does not. The policy continues to be aimed at holding mine managers accountable for what goes wrong. The policy of centralised control would mean that the BHP hierarchy, perhaps even its executives in Melbourne, might truly be held accountable. In the absence of some kind of organisational restructuring along these lines, the claim with which this chapter began, namely that BHP accepts accountability for the disaster, seems largely devoid of meaning.

10

THE 'SAFETY PAYS' ARGUMENT

Government agencies regularly use the argument that 'safety pays' as a way of motivating employers to attend to occupational health and safety. The assumption is that if employers can be made to realise that expenditure on safety is a good investment they will have a financial incentive to improve workplace safety. In so far as employers embrace this view, the need for government agencies to enforce OHS law will be minimised.

The Moura disaster provides us with an opportunity to examine the value of this strategy. We saw in Chapter 8 that safety was sacrificed to production. According to the 'safety pays' argument, a more far-sighted and rational approach would have seen managers paying greater attention to the question of catastrophic risk. The ultimate questions to which this chapter seeks answers are these: does safety pay in contexts such as Moura and does this argument have the potential to provide decision makers with the motivation to attend conscientiously to catastrophic risk?

THE MACRO APPROACH

Advocates of the argument that safety pays often present national figures to support their case. They calculate the numbers of injuries nationwide and from this produce an estimate of the total cost borne by employers, the workers concerned, their families and governments. In Australia estimates range from $20 billion (IC 1995:94) to $37 billion (NOHSC 1994). These figures are quite beyond most people's comprehension. To put them in context, the Industry Commission (IC) notes that the figure of $20 billion is equivalent to 5 per cent of GDP, that is, about five per cent of the country's total production of goods and services (IC 1995:94). It notes too that the cost averages about $27 000 for every injury, fatality or illness (1995:94).

Purse (1997) seeks to put the figure of $20 billion in context by noting that it is greater than the output of entire industries, more for instance than the net output of the agricultural sector and more than is produced by Australia's mining industry. He goes on:

> comparisons with Commonwealth government budget outlays further illustrate the magnitude of the problem. In 1992–93 allocations to the largest portfolio areas in the Commonwealth budget were all less than the losses attributable to work-related injury and death. Neither the defence budget, Medicare, higher education expenditure nor the financing of the social security system were as costly as the failure to manage workplace health and safety.

These comparisons serve well to give us a sense of the magnitude of the sums involved. But what are the implications? Does it follow that if we were able to eliminate occupational injury and disease the country would be $20 billion better off? The answer is no, as those who make these calculations are careful to point out. The reason is that many of the costs involve benefits to others. Consider this. The cost to the employer of putting other workers on overtime to cover for the absence of injured workers is a benefit to those doing the overtime; the cost of replacing or repairing damaged equipment represents increased income to the firm which carries out the repairs or provides the replacement; the cost of medical, rehabilitation and retraining services represents increased revenue to these service providers and so on. In short, many of the costs of injury and illness are at the same time gains for someone else. Thus the net costs of injury and illness to the country as a whole is far less than the figure of $20 billion.

We can get some idea of just how misleading the total cost figure is from the following calculation. Suppose we were able to reduce injury and disease by ten per cent. This would reduce the total cost figure by some $2 billion. But the IC estimates that this would produce a *net* benefit to the nation of only $0.34 billion

(1995:531). It is clear therefore that *aggregated* national figures on costs of injury and illness are thoroughly misleading.

In any case, if the aim is to motivate action, it is more useful to identify costs to particular parties. An obvious party is the federal government and the IC calculates that the cost of workplace injury and illness to the Commonwealth budget is $3.5 billion, most of which is incurred in the social security budget (IC 1995:107). This kind of estimate provides a very direct motive for action. The federal government stands to save billions if it can substantially reduce the workplace toll. In contrast, the estimate of $20 billion for the nation as a whole provides no one in particular with a reason to act.

There is an important lesson here. Any attempt to argue that safety pays must specify for whom. Unless we can identify a relevant decision maker for whom safety pays, the argument has no capacity to motivate action to reduce injury and illness. This idea will emerge as a critical theme later in this chapter.

Since it is employers who are in the best position to control workplace harm, it makes sense to ask what the costs are to them. The Industry Commission estimates that employers incur about 30 per cent of the total cost, the rest being borne by the worker and the community (IC 1995:102). Moreover, for more severe harm the proportion of cost paid by employers declines. 'Hence [says the IC] financial incentives for employers to prevent injuries are inadequate particularly for serious incidents . . . The low share of costs borne by employers for severe workplace incidents proves a strong rationale for regulation' (IC 1995:102). Ironically, the IC is arguing that in this matter safety very definitely does not pay.

The IC is somewhat off track here. There are good arguments for regulation, but this is not one of them. The calculation that rational employers make is not what proportion of the total cost they bear, but whether the cost of prevention outweighs the cost borne by the employer, regardless of what proportion of the total cost that may be. The employer might well bear 100 per cent of the costs of injury and still make the judgment that these costs are less than the cost of the investment necessary to make the workplace completely safe.

Nevertheless, the IC's argument on this point is at least a recognition that the total cost approach is devoid of implications and that cost arguments only develop potential leverage when we focus on who is paying those costs.

Costs at the enterprise level

The British government's Health and Safety Executive (HSE) has developed this argument further in a major study of the costs to employers of workplace accidents in five organisations: a creamery, a construction site, a North Sea oil platform, a transport company and a National Health Service hospital (details from Cutler & James 1995). The study included incidents that produced either injury, or economic loss without injury. Thus, while not just a study of the costs of injury, its findings are potentially relevant to the question of whether safety pays. The study was conducted over a period of three to four months and collected data on a total of over 600 accidents. Average cost per accident ranged from 761 pounds in the case of the construction site to 2951 pounds in the case of the oil platform. On an annual basis the costs represented 8.5% of the tender price of the construction organisation, 1.4% of the creamery's operating costs, 37% of the transport company's profits, 14% of the potential output of the oil platform and 5% of the annual running costs of the hospital. Cutler and James note that 'at a superficial level these figures appear to lend considerable support to the HSE's argument that safety pays and hence that there is no inherent conflict between expenditure on safety and organisational profitability' (1995:3).

This conclusion does not follow, however, as the authors point out. The HSE's figures are quite indeterminate; an enterprise manager faced with such figures needs to take account of the costs that would be incurred in seeking to prevent these accidents before being able to make a rational financial calculation.

It seems probable that where particular kinds of injury are occurring with relatively high frequency, cost-benefit calculations will lead rational employers to invest in safety. Consider, for instance, the epidemic of RSI that occurred in the 1980s in Australia and cost many employers large sums in terms of compensation premiums, if nothing else. It was economically rational for such employers to spend money on a variety of preventive measures such as limiting hours spent doing repetitive work, installing ergonomically sound furniture, designing ergonomically optimal tools and so on. But it cannot be assumed without evidence that it will always be rational for employers to act in this way, or even that employers will act rationally in such matters.

One reason why managers sometimes fail to make appropriate safety investments is misleading accountancy practices.

Johnson and Kaplan (1987) point out that accounting systems are often designed to compare income and expenditure over a short period, say a year, to provide some measure of profitability during that period. These calculations assist money managers to calculate 'return on investment' and thus to decide where financial resources may be optimally deployed. But looking at cost effectiveness over one year is misleading. The costs of the relatively few accidents occurring in one year may not be such as to justify the large capital outlay necessary to prevent those injuries. However, the savings over many years will be many times larger. If the expenditure could be deducted against budgets over that much longer period instead of being counted as an expense in the year of purchase, the expenditure might seem far more cost-effective.

To go back to the RSI example, the payoff in one year from re-equipping the office with ergonomically designed furniture may not appear to warrant the expenditure. But the savings over many years may far exceed this one-off cost. This is only one aspect of Johnson and Kaplan's case. They show in great detail how accounting practices are preventing managers from making rational decisions in relation not just to safety but also to a whole range of production and investment decisions, and they call for the introduction of new practices to aid managers in rational decision making. The point here is that such practices could be expected to reveal that expenditure on safety is more cost-effective than is currently realised.

IS IT ECONOMICALLY RATIONAL TO PREVENT DISASTERS?

Our particular interest in this book is events that are relatively rare but catastrophic. Let us consider whether it makes economic sense to spend money to prevent disasters. Intuitively the answer is yes. Disasters may cost anywhere from millions to billions of dollars. (The Piper Alpha disaster is estimated to have cost more than $4 billion, Wright 1994:95). Presumably it would pay to spend large sums of money to avoid such losses. Let us take a particular example to illustrate the point.

In 1987 a car ferry, the *Herald of Free Enterprise*, left Zeebrugge in Belgium on the cross-channel run to Dover. As it left port it capsized and 192 people were drowned. The vessel was a roll-on-roll-off design, allowing vehicles to drive on one end and off at the other. The ship's bosun had failed to close the watertight bow doors before departure and as the ship left harbour the vehicle deck filled with water and the vessel foundered within four minutes. The losses to P&O, the company that owned the ferry, were presumably substantial.

A number of factors contributed to this disaster, among them the fact that the ship's captain had no reliable way of knowing whether the bow doors were closed. Ferry masters had been concerned about this and had written to senior management asking that warning lights be fitted on the bridge of ferries in the fleet. Such lights would not have involved a major investment but would have substantially reduced the risk of sailing with the bow doors open. From this point of view the safety investment would appear to make good sense.

There is more to it than this, however. In making a decision about whether a safety investment is worthwhile, some account needs to be taken of the likelihood of the disaster the investment is designed to prevent. If the probability of disaster is vanishingly small, it may not be rational to make further safety investments. But in the Zeebrugge case the risk was not vanishingly small. Prior to this disaster there had been five recorded instances of company ships proceeding to sea with open bow doors. Each of these incidents could easily have been a disaster.

Strictly speaking, in order to decide whether a particular safety investment made sense economically, one would need to be able to make some quantitative estimate of the likelihood of the disaster and include this in the cost calculation. A simple way of thinking about this is to consider what kind of premium a prudent insurance company would charge to insure a roll-on-roll-off vessel against capsize, *knowing that there had been five previous sailings with open doors.* Presumably, the figure would be very considerable. On the other hand, a rational insurer would be willing to charge a much smaller premium for a vessel with warning lights installed. The difference would indicate whether or not it made sense from a strictly financial point of view for the company to install the warning lights on the bridge. Viewed in this way, it seems highly likely that this particular safety investment would be judged to make economic sense.

The fiery destruction of the space shuttle, *Challenger*, is one disaster where the risks *were* quantitatively assessed—unfortunately, after the event. The conclusion was that there was a 13 per cent probability of catastrophic O-ring failure at the low temperature at which the launch took place (Dalal et al. 1989). Putting this another way, if eight launches occurred under these conditions, one could be expected to end in disaster. Clearly, our mythical insurance company, knowing this probability, would have required such a high premium to insure a launch under these weather conditions that, regardless of the costs incurred by deferring the launch until the weather was more favourable, it would have been worth NASA's while to do so, because of the much lower insurance premium that would then have to be paid.

Institutionalised irrationality?

To return to the *Herald of Free Enterprise*, the question that arises then is why, given the economics of the situation, warning lights were not installed as the ferry masters requested. When one master wrote a memo to senior management two years before the disaster, the memo was circulated to various members of senior management for comment. These comments are preserved for all to see. One wrote: 'Do they need an indicator to tell them whether the deck storekeeper is awake and sober? My goodness!' Another wrote: 'Nice, but don't we already pay someone!' (quoted in Cutler & James 1995). The request was rejected. Critical decision makers, in other words, failed to apply their minds to the matter in any rational way and certainly failed to weigh up the economic arguments.

Other disaster analyses come to the same conclusion—that, although safety pays, organisations frequently act in ways that are quite irrational from this point of view. In his analysis of why safety systems on Piper Alpha were so inadequate, with such economically disastrous results, Wright (1994) comes to the conclusion that Piper exhibited a form of 'institutionalised irrationality'.

Bounded rationality

We cannot leave the matter here, however (and Wright doesn't). To describe behaviour as irrational is no explanation. It is simply an admission that the behaviour is inexplicable in terms of the concepts to hand. A concept with the potential to advance our understanding a step further is Simon's idea of bounded rationality (1957). He argues that decision makers do not have access to all relevant information and could not process it even if they did. They are thus not in a position to make decisions that are rational in any global sense. Instead, they evaluate the information readily to hand and choose from among the options already known to them. Decision making is rational within these bounds.

A slightly different way of conceptualising the sometimes apparently irrational behaviour of decision makers is Etzioni's concept of disjointed incrementalism:

> The decision maker, rather than attempting a comprehensive survey of all alternatives, focuses instead only on those policies that differ incrementally from existing policies. Only a relatively small number of policy alternatives are considered. For each policy alternative, only a restricted number of important consequences are evaluated. Thus there is no one decision or right solution but a never ending series of attacks on the issues at hand through serial analysis and evaluation. The term 'disjointed incrementalism' is used to emphasise the lack of major direction, policy or course-setting capacity. The incrementalist fate . . . is to stumble through history, putting one drunken foot in front of the other. (Etzioni 1988:126; Etzioni's attributions omitted)

Whereas Simon's image is one of a rational decision maker operating within bounds, for Etzioni the bounds are so limited that the behaviour is about as sensible as that of a drunk; to use another of Etzioni's phrases, it amounts to no more than 'muddling through' (1998:126).

The idea that decision making is really a process of muddling through rather than a rational evaluation of all alternatives has been well developed in relation to the behaviour of firms. Research has shown that management decisions typically do not result in optimal performance or profit maximisation. Managerial attention can be focused on only one or a small number of things at a time and cannot encompass all the factors that would need to be taken into account to optimise performance. In fact, attention is focused on matters that are causing the greatest concern or where performance is falling furthest below expectations. Managers often find themselves rushing from one thing to another, 'putting out fires', rather than rationally calculating how best to achieve the firm's goals (Scholz & Grey 1990:286).

The best-known statement of the view that firms behave suboptimally is Cyert and March's *A Behavioural Theory of the Firm*.

> In neo-classical theories of the firm, organisations identify, choose, and implement optimal alternatives. In behavioural theories, organisations simplify the decision problem in a number of ways. They set targets and look for alternatives to satisfy those targets, rather than try to find the best imaginable solution. They allocate attention by monitoring performance with respect to targets. They attend to goals sequentially, rather than simultaneously. They follow rules-of-thumb and standard operating procedures (which simplify and routinise decision making but do not necessarily result in optimal outcomes). (1992:214–15)

The notions of bounded rationality, disjointed incrementalism, muddling through and putting out fires provide considerable insight into the apparent failures of rationality in the matter of catastrophic risks. But various puzzles remain. Consider the behaviour of the P&O management. Despite the fact that their attention was drawn to the risk of capsize, they failed to respond effectively. One might have expected that no matter how limited their information or their capacity to process it, once the issue had been raised they would have responded in a more rational way. Does irrationality have a part to play in the explanation of disaster, after all? If we are to avoid this conclusion we need once again to change our conceptual approach.

For whom does safety pay?

To understand why the 'safety pays' argument appears at times to be so ineffective we need to address again the question: for whom does safety pay? An analysis of the Bhopal disaster reveals the crucial significance of this question.

In 1984 toxic gas leaked from a chemical plant owned by Union Carbide in Bhopal in India. It drifted over nearby residential areas, killing at least 2000 people and injuring more than 200 000 (Kletz 1994:95), the worst industrial disaster in history (Lepkowski 1994). Union Carbide finally agreed to a settlement of $470 million to be paid to victims. If account is taken of lost production and company time and energy spent dealing with the matter, the total cost would be considerably higher. To put these losses in perspective, Carbide's profit in 1982, two years before the disaster, was $310 million (Lepkowski 1994:27). In short, the cost of compensation alone, when eventually paid, was equivalent to one-and-a-half times the company's entire profit for 1982. In anticipation of these losses, the share price dropped from $60 to

$30 following the disaster, effectively halving the market valuation of the company. In other words, almost overnight shareholders saw the value of their investments cut in half. These were dramatic losses by any standard.

The sequel to the disaster is quite fascinating. Union Carbide found itself subject to a hostile takeover attempt by GAF Corporation which deepened the financial crisis. To defeat the takeover and extricate itself from the crisis, Carbide borrowed $3 billion and paid its top officers a total of $28 million in so-called 'golden parachutes' as insurance against further takeover attempts. To pay off its debts the company was forced progressively to sell some of its most profitable component businesses. The result, after all this divestment, was a much smaller company, one-sixth the size of its former self. But it was also more focused on its core business; and *more profitable than ever*. In 1988 the company made record earnings per share and the year was reported as 'the best in the 71-year history of Union Carbide' (quoted in Lepkowski 1994:30). All this has led one student of Union Carbide's fortunes to the following conclusion:

> Clearly, by any objective measure, Union Carbide Corporation *and its managers* benefited from the Bhopal incident as did Union Carbide of India Ltd. They were politically able to close a burdensome plant, take aggressive actions to restructure both companies, and *enhance management benefits* . . . It is ironic that a disaster such as Bhopal [would] leave its victims devastated and other corporate stakeholders better off (quoted in Lepkowski 1994:30, emphasis added).

The point to note is that, despite the costs involved and despite the losses to those who owned shares at the time of the disaster, top management actually benefited financially. Top managers are the people whose decisions are crucial to safety. Yet, as it turned out, they had no financial interest in safety. On the contrary, as a group their inattention to safety paid off handsomely in this instance.

None of this is to deny that there may have been psychic or emotional costs paid by top managers. The chairman of Union Carbide, in particular, lived for years with the possibility that he might be extradited to India to stand trial. But this involves a very different set of motives, addressed in Chapter 11. The 'safety pays' argument is a strictly financial argument and that is what is under consideration at this point.

The Bhopal case provides evidence that, no matter how costly a safety failure may seem for the company as a whole, top managers themselves may be shielded from any untoward financial consequences. This provides an important insight into why—in relation to safety matters—management acts in ways that seem irrational to outside observers. The relative indifference to safety displayed by senior officers at P&O, which appeared quite irrational from the point of view of any overall cost–benefit analysis, may well have been quite rational in terms of the personal cost–benefit calculations implicitly made by these decision makers. Cutler and James note that the tragedy did not damage the commercial prospects of P&O's ferry business and that cross-channel ferry traffic expanded substantially in the following years. It is a reasonable hypothesis that P&O's top management suffered no financial ill effects from the tragedy.

As for Piper Alpha, in the aftermath of the disaster the operating company, Occidental, withdrew entirely from the North Sea and became a smaller company. But, as the Bhopal case makes clear, it cannot be assumed that the company's top decision makers suffered any corresponding financial consequences.

To sum up, attention to the question of for whom safety pays, and in particular whether it pays for relevant decision makers, enables us to move beyond explanations in term of irrationality and to make more informed judgment about whether appeals to the notion that safety pays are likely to be as influential as government authorities presume they will.

MOURA

Against this background, let us now address the question of whether safety might have paid at Moura and the extent to which this proposition had the potential to motivate relevant decision makers. Consider first whether or not safety paid from a rational economic point of view. Instituting a regime of careful monitoring of spontaneous combustion would have been relatively inexpensive. All the necessary equipment was already on site and all that would have been required was for someone with the necessary expertise to carry out the task conscientiously. Moreover, the risks of explosion at Moura were statistically high. There had been two explosions with loss of life in immediately adjacent mines in 20 years and various other confirmed and suspected heatings. From this point of view, playing it safe made good economic sense.

So why did this financial argument not prevail? To answer this we need again to focus on the question of for whom safety might pay. Consider, first, the mine's ultimate owner, BHP. What financial impact would an explosion at Moura have on this company?

Moura underground mine was a tiny part of the BHP empire. To put it in perspective, BHP operates mines around the world producing a variety of minerals as well as coal; it has a petroleum division that operates worldwide and, of course, it produces steel and steel products in various parts of Australia. Moura underground mine contributed less than two per cent of all the coal exported in the previous year from Queensland coal mines operated by BHP. Even at the Moura site itself the underground mine contributed only 15 per cent of the total production, the rest coming from an open-cut operation. According to one Sydney resource analyst, 'as far as BHP is concerned [the explosion] might be compared with a week of bad rain'. Another said that the disaster would not have 'any impact on the market; it will have nil impact on profit and minimal impact on revenue'. In seeking to evaluate the economic impact of the loss of the mine on BHP it is also useful to recall that in the year prior to the explosion BHP reported a profit of 1.28 *thousand million* dollars. (For details above see *Courier Mail*, 10/8/94, p. 16.)

In short, if BHP loses a mine from time to time, this is of no financial consequence as far as shareholders are concerned. Nor, therefore, does it have any financial impact on the most senior executives of the company. They are thus under no financial pressure to take a more active interest in mine safety. This is not to say that senior managers are unconcerned about safety; simply that financial considerations play an insignificant role in generating concern.

Consider next the economic pressures at work on the management of Moura underground mine. As already noted, the future of the mine was in doubt and Moura was under pressure to produce as fast as possible. The fate of the mine and the livelihood of its workers depended on it. Even managers of the mine could conceivably be out of a job if the mine closed. Did safety pay for these managers? Certainly, attention to the LTIFR paid off, as we noted in an earlier chapter, but what of attention to the possibility of spontaneous combustion? The most serious conceivable economic consequence mine managers faced if an explosion occurred was mine closure and the possibility of job loss, the same outcome that might ensue if production levels proved inadequate. Moreover, from their point of view the prospect of mine closure due to production shortfalls was more immediate than mine closure due to an explosion. In short, from a purely economic point of view, and given that their capacity to manage was not unlimited, it made economic sense to focus their attention on production.

This is clearly a highly constrained conclusion. Economic costs are not the only costs involved. Managers are not just economic calculators; they are also human beings capable of feelings of empathy, sorrow, shame and guilt. The psychic costs of a disaster would far outweigh the psychic costs of mine closure due to production shortfall. In terms of psychic costs it would be easy to make the argument that safety pays. But that is not the point. What is at issue here is whether safety pays for mine managers in a purely economic sense and, stating the matter as cautiously as possible, it is far from clear that it does.

The upshot of this analysis is that, for BHP, regardless of the point in the company hierarchy at which decisions are made, economic considerations did not seem to operate to provide safety incentives in relation to catastrophic risks. If economic considerations are to play any role in the prevention of future underground coal mine explosions, BHP will need to redesign its remuneration structure for managers to provide incentives to focus on such risks. What is needed is some quantitative indicator of how well managers are managing catastrophic risk. Bonuses can then be paid on the basis of this indicator. But there are no simple and reliable indicators of how well managers are managing in this matter. Unless indicators can be developed, the 'safety pays' argument will remain an ineffective way of encouraging managers to focus on catastrophic risk.

Conclusion

This chapter has uncovered a number of problems with the 'safety pays' argument as a means of controlling catastrophic risk. Even though it would appear to be entirely rational from an economic point of view to devote considerable resources to minimising the risks of disaster, organisations far too often fail to do so. In so doing they act in apparently irrational ways. Part of the explanation for this lies in the inability of organisations and their managers to act in fully rational ways; all that can be hoped for is rationality within certain bounds. But in addition we must take seriously the question of for whom safety pays. It turns out that all too often top managers have no financial incentive to attend to catastrophic risk. In short, at a personal level the 'safety pays' argument simply does not have the capacity to motivate top managers to attend to safety. Unless the structure of remuneration for managers can be changed to provide the necessary motivation to focus on catastrophic risk, 'safety pays' will be an ineffective slogan.

We might go further than this and conclude that the message that government agencies should be seeking to convey to managers is not that safety pays *financially* but that it pays in *psychic* terms. Suppose managers were regularly asked to consider the following questions:

- *How would you feel if the mine you are managing exploded?*
- *How would you feel if eleven men were killed as a result of your failure to attend adequately to questions of safety?*
- *How would you feel if you were confronted by miners' widows and children asking why their loved ones had to die?*
- *How would you feel if subject to legal action?*

In the matter of disaster prevention, invoking these motives is likely to be more effective than any argument about financial cost.

11

THE REGULATORY SYSTEM

What type of regulatory system is most likely to minimise risk of future coal mine explosions? This is the question addressed in this chapter. It is a timely question since the system of government regulations aimed at ensuring coal mine safety in Australia is currently changing. This chapter is thus in part a commentary on those changes.

Coal mine safety regulations are a state responsibility in Australia. They were first enacted last century and progressively amended as technology advanced. They have, traditionally, been highly prescriptive in nature, specifying in great detail how safety is to be achieved, and based on the assumption that government is in the best position to determine how miners' health and safety is to be protected. The role of inspectors in this regime is to ensure that the rules are being complied with.

This was the approach to regulating occupational health and safety for industry generally, in Australia and the UK, until the 1970s. But the 1972 Robens Report in the UK triggered a broad trend away from detailed prescription towards what is often described as self-regulation. This concept is now on the agenda in the area of coal mine safety regulation as well.

As might be expected, self-regulation does not, or at least should not mean that governments abandon the field. Self-regulation is not just another word for deregulation. Rather, it involves a new style of legislation that sets up a framework in terms of which health and safety is to be managed. It is for this reason sometimes described as enabling legislation.

One aspect of this enablement is to provide workers with various rights to be involved in decisions about safety, rights to information and training and so on. All Robens-inspired legislation specifies such rights. This is an important but sometimes overlooked aspect of self-regulation.

Moreover, while the new-style legislation frees employers from the legal requirement to abide by specific rules, it requires employers to meet standards variously described as performance or goal setting standards (Brooks 1994; Woolfson & Beck 1995). For example, most such legislation requires employers to provide a safe and healthy workplace, so far as is reasonably practicable. It is then up to the employer to decide how to do this. Governments may write codes of practice to provide employers with guidance on how they can achieve this goal, but employers are under no obligation to follow these particular guidance rules as long as they have in place a system that is equally effective. The simplest strategy for an employer who wishes to play it safe is to regard these codes as mandatory, but for employers who wish to develop their own procedures the scope is there.

This general requirement that employers secure the health and safety of their employees so far as is reasonably practicable is known as the duty of care. Failure to exercise this duty of care can lead to substantial penalties, much higher than were possible under the old-style legislation. Under the new-style legislation employees also owe a duty of care to each other and themselves.

Critics have pointed to drawbacks of the self-regulatory approach in certain contexts. One such context is small business, which here means any business too small to employ a safety specialist. Despite all the rhetoric about wanting relief from the burden of regulation, small businesses do not like being told that they have a duty to maintain a safe workplace but that it is up to them to decide how to achieve this. They want to be told precisely what to do so that they can do it and get on with the job, secure in the knowledge that they have complied with their obligations (ACIL 1997:66).

A second context in which it has been shown that self-regulation is deficient is where the workforce is not unionised. In such environments workers do not have the confidence to become involved in safety matters and there may be no one in a position to put pressure on employers to take their obligations seriously (Dawson et al. 1988).

Coal mining in Australia does not involve small businesses and is highly unionised. Neither of these difficulties therefore applies. Is the move to self-regulation in the coal industry therefore appropriate?

PRESCRIPTION OR SELF-REGULATION?

Moves towards self-regulation of the coal mining industry in Queensland were suspended following the Moura explosion, pending the outcome of the inquiry. BHP took the opportunity to argue at the inquiry that the explosion was evidence that prescriptive regulation had failed and that the authorities should press on with self-regulation. The view of the mining union was that self-regulation was likely to be even less effective than prescriptive regulation and that the proposals for self-regulation should be dropped. The inquiry panel came up with a compromise proposal in relation to this issue. It suggested that 'high probability, low consequence matters [which give rise to the run-of-the-mill lost time injuries] might suitably be addressed by self-regulation but that low probability, high consequence matters [such as explosions] should remain subject to prescriptive legislation' (Windridge 1996:65).

It is interesting to note that a view entirely contrary to that of the inquiry panel has been expressed in the UK. Glen (1993) has suggested that, in hazardous industries where disasters are a possibility, the operating companies themselves can be expected to have a keen interest in disaster prevention and that in these circumstances self-regulation can be expected to work best. While the evidence about disasters presented in this book does not confirm this view, the view itself is further evidence that the debate is far from settled.

TRANSCENDING THE ISSUE OF PRESCRIPTION VERSUS SELF-REGULATION

There are problems, however, with this formulation of the issue as one of prescription versus self-regulation. Although the changes occurring in coal mine safety legislation are frequently described as a move from prescription to self-regulation, this is a quite misleading description. To start with, there are already elements of self-regulation within the existing 'prescriptive' acts. For instance, the legislation requires that management make its own roof support rules in each mine, specifying whether support is needed and, if so, the type of support and at what intervals. These details are not laid down in the legislation. What is required by law is that the roof support rules be submitted to the inspectorate for approval. Existing legislation, then, already involves a departure from any simple notion of prescription. The regulation of roof support involves significant elements of self-regulation.

Consider, next, one of the most important changes occurring in the regulatory regime. Inspectorates in both NSW and Queensland now require that mines develop plans for the management of specified catastrophic hazards. The details of these plans are not specified by regulation or by inspectorates and are ultimately for mine management to decide. Safety management plans are often seen as one of the central features of the self-regulatory approach.

But, although the inspectorates do not write the plans, they have drawn up a comprehensive set of requirements as to what must be included in the plans. What is prescribed is not the detailed content but the matters that must be considered and the processes that must be undertaken. There is, in short, a high degree of prescription in the new approach.

Here, for instance, is a very small part of the NSW requirements for management plans for dealing with spontaneous combustion. It is the section that refers most directly to the need for clear procedures for responding to CO gas alarms and other tell-tale signs (NSWDMR, October 1995).

> . . . the mine must have documented evaluation/decision processes for the following:
>
> - the setting of trigger levels or conditions for spontaneous combustion indicators in use at the mine which might result in the activation of predetermined decision processes; . . .
>
>   *Note*: indicators in use may include observations other than those based on gas analysis and interpretation [e.g. smells and heat hazes].
>
> - the development and implementation of predetermined responses (action plans) to defined triggers indicative of spontaneous combustion . . .
>
> These documented evaluation/decision processes must identify who must be involved in each process, who has authority for the decision(s) and the criteria to which the decisions are to be made . . .

Note that this small passage contains the word 'must' three times. The document contains a large number of such so-called 'mandatory elements'. Indeed, on just one of its sixteen pages, the word 'must' appears thirteen times. It is clear therefore that this new self-regulatory approach is highly prescriptive.

Finally, it is notable that the draft regulations drawn up in NSW in 1997 retain a great deal of prescription although they also introduce, for mine managers and workers, the concept of duty of care, a central feature of the self-regulatory approach. These regulations demonstrate that it is not an either/or choice and that it is possible to combine the best of both approaches.

In summary, coal mine legislation has always had elements of both prescription and self-regulation. What is now occurring is that new elements are being introduced into the regulatory regime, elements that are often described as self-regulatory. These additions do not signal any wholesale move away from prescription. They demonstrate in fact that the two approaches can happily coexist.

LIGHTENING THE BURDEN OF REGULATION?

One of the reasons sometimes advanced in favour of self-regulation is that it will lighten the alleged burden that business currently bears. This is often the hope of large employers, particularly those who have operations in more than one state and are subject to more than one regulatory regime. But consider the compliance task facing management in relation to the new safety management plans. It is two-fold: first, to ensure that the plans conform to the requirements specified by the inspectorates; and secondly to ensure that they are scrupulously observed.

This compliance task does not seem to be any less burdensome than under a so-called prescriptive regime. On reflection, this is not surprising. It is clear that companies must expend considerable resources to ensure that catastrophic hazards are properly controlled. There is no way around this, regardless of what kind of regulatory regime is in force.

A REDUCED ROLE FOR THE INSPECTORATE?

Another reason that is sometimes advanced for self-regulation is that it shifts responsibility for safety back to industry, thereby reducing the need for inspectorates and enforcement. This is seen as desirable, given current beliefs about the virtues of small government.

But the new regime does not herald a reduced role for the inspectorate—rather, a new role. Not only must inspectors ensure that mine safety management plans comply with the new prescriptions; they must carry out a variety of audits to ensure that the plans are working as intended. This is a time-consuming activity and the indications are that inspectorates will need increased resourcing if they are to be able to carry out this new role effectively. There is already evidence that companies are developing their own plans without reference to the requirements laid down by the inspectorates; the process of bringing these companies into compliance with the inspectorate's standards may not be easy.

If the auditing process is to be effective inspectors must be in a position to issue legal notices requiring companies to comply. Such notices are routinely used by the general OHS inspectorates but not by mining inspectorates. These notices should land on the desk of the company's chief executive officer, not simply that of the mine manager, so that the most senior company officials are encouraged to take an interest in compliance. When inspectors discover flagrant disregard for plans which on paper are supposed to be in operation, they should be in a position to prosecute for breach of duty of care. Thus the new approach does not in principle require any less enforcement activity on the part of the inspectorate.

Companies sometimes argue approvingly that the move to auditing is a move away from enforcement. What I am suggesting is that this is not necessarily the case. Companies that are not managing safety according to an agreed plan are in breach of their duty of care, and it is vital that regulatory agencies be able to take strong enforcement action in such cases.

THE FAILURE OF ENFORCEMENT

It is frequently suggested that self-regulation is necessary because prescription has failed. As we saw above, BHP argued at the inquiry that the Moura explosion was evidence of the failure of prescriptive regulation. There is an important counter view to be considered, namely that prescriptive regulation has never been given a chance. The argument here is that coal mine inspectorates have been either not able or not willing to use their powers of enforcement and as a result the prescriptive regime has not been as effective as it might have been.

Consider this statement by a recent NSW government inquiry (ACIL 1997:70):

> . . . the Review is not as convinced as are some stakeholders that the present rates of serious injuries and fatalities are an indication that the current legislative approach has failed. It could equally be argued, given the number of accidents which relate to breaches . . . that present accident rates are primarily an outcome of inadequate enforcement of existing legislative provisions.

Coal mine inspectorates in Australia almost never prosecute for violations they discover. I was told that the Queensland inspectorate has launched only two prosecutions in sixteen years, one of which was for repeated failure to comply with electrical rules. The policy is to seek compliance by asking, and only if defiance is encountered is prosecution contemplated. Not even deliberate violations are prosecuted. On one occasion, an electric motor that was causing annoyance by overheating and automatically cutting out was illegally modified so that the cut-out mechanism did not work. It overheated again, continued to operate and caused some coal to catch fire. This was a serious offence and the offender was spoken to severely. But an inspector's recommendation to prosecute was not pursued by his superiors.

The policy of non-prosecution was dramatically illustrated following an explosion in a NSW coal mine in 1995 which, by sheer good luck, killed no one, even though there were 30 men underground at the time. The inspectorate investigated the circumstances in great detail and produced a report in which it identified numerous violations of safety regulations. However, no prosecution was mounted. The inspectorate announced that it had done the investigation and published the report so that other mines might draw whatever lessons they could from it (NSWDMR 1996).

This policy of non-prosecution is matched by a non-confrontational inspection style. The manager of Moura gave evidence of what happens. The inspector generally announced his intention to visit the mine. When he arrived the manager sat him down for a cup of tea and asked what he wanted to see. He then accompanied him underground on the inspection. The manager had nothing to hide, he said, and took the inspector anywhere he wanted to go. Any suggestions or requests made by the inspector would be followed up. It should be noted that there is nothing unusual about the fact that inspections at Moura were pre-announced. This is routine in the industry. The recent NSW government report was critical of this practice and suggested that at least some visits should be unannounced so as to minimise the risk that mine managers might be deceiving inspectors (ACIL 1997:60).

The very cordial relations between mine managers and inspectors, coupled with the reluctance to prosecute, suggests that the inspectorates are somehow constrained in the way they do their job. The following dilemma in which one Australian mining inspectorate finds itself provides direct evidence of this constraint.

In mines where methane has never been found, managers do not wish to go to the expense of equipping their mining machines with methane gas monitors, at a cost of approximately $50 000 each. The mines inspectorate believes that best practice requires that all machines be fitted with such monitors in case, as mining progresses into new areas, gas is encountered unexpectedly. The inspectorate has enormous powers theoretically, and could simply order that this be done. It does not do so for fear that mine owners might complain to the minister about the inspectorate forcing them to spend money unnecessarily. It so happened that one of these presumed non-gassy mines struck gas. The deputy had just done a perfunctory check (perfunctory because he was not expecting anything) and had missed the gas which had apparently been building up. An explosion occurred and two men were badly burned. There could be no more graphic demonstration of the wisdom of the inspectorate's position. Nevertheless, other 'non-gassy' mines continue to refuse to install the appropriate monitors and the inspectorate still feels unable to use its powers to compel them. Clearly, this inspectorate is shackled by the political context in which it operates.

How is it that coal mine inspectorates have been shackled in this way? For enlightenment I turn again to the Piper Alpha explosion and the findings of the commission of inquiry. Safety in the North Sea was the responsibility of an inspectorate located within the Department of Energy. This was the department with overall responsibility for the development of petroleum resources in the North Sea. Its job was to facilitate the activities of the exploration and production companies involved. The inquiry found that the role of the safety inspectorate was compromised by its location and that the Department of Energy would inevitably be more sensitive to the needs and wishes of the operating companies than was desirable from a safety point of view. The inquiry therefore recommended that henceforth safety in the North Sea be the responsibility of Britain's Health and Safety Executive, a government agency that was independent of any particular industry and whose sole function was the protection of worker health and safety.

The coal mine inspectorates in NSW and Queensland suffer from a similar structural problem. They are both located in mining departments whose primary role is to assist industry to develop the state's resources. They are inevitably compromised by this location. Both coal inspectorates have recently been amalgamated with the metalliferous mine inspectorates in the same departments. But if the logic of the Piper Alpha inquiry is applied, they should be relocated in the generalist occupational health and safety inspectorates in these two states.

It seems likely that, until these structural arrangements are changed, the enforcement function of the inspectorate will remain compromised. This is particularly worrying as the industry moves into the new era of safety management plans which depend ultimately for their effectiveness on rigorous auditing by inspectorates. As argued above, if inspectorates are not able to compel mines to adopt satisfactory safety plans and to compel compliance with them, the whole strategy for the control of major hazards will be undermined.

The importance of duty of care

I turn now to duty of care, a concept with the potential to increase dramatically the attention that coal mining companies pay to safety. It must be said at the outset, however, that this potential will only be realised if inspectorates are enabled to take a tougher line on enforcement.

Traditional coal mine legislation imposes obligations on people at the mine site up to and including mine managers. However, under NSW law the mine owner is largely shielded from any liability (section 164(2)). Under the Queensland legislation in force at the time of writing the corporation that owns the mine has limited protection against liability (section 104(2)) and its directors have total protection (section 104(7)).

New Queensland legislation (in draft form at time of writing) has a much greater focus on the obligations of the operating company in that, among other things, it requires the company to provide a safe place of work. Moreover, the maximum penalty that could be imposed on the operator in the event of a violation of this obligation is $60 000. Under the old legislation the maximum penalty is $200. The new legislation will bring the situation of mine operators roughly into line with operating companies in other industries, in theory. But in practice nothing will change unless the mines inspectorate is enabled to prosecute operating companies for violations of the general duty of care as is now routinely done by the inspectorate in the general occupational health and safety arena.

In NSW, as well as being subject to the *Coal Mines Regulation Act*, the coal industry is formally subject to the general OHS Act with its requirement that employing companies exercise duty of care. Operating companies are therefore potentially liable. In practice, though, the general OHS inspectorate leaves mining to the mining inspectorate with its de facto policy of non-prosecution. Coal mine operators in NSW have thus in practice been immune from prosecution.

There are good reasons for holding companies liable for the failure of duty of care—that is, the failure to maintain a safe working environment at the mine site. In Chapter 9 it was argued that responsibility for the management of catastrophic hazards should be located at the company level and not decentralised to the mine site. Making companies legally liable would encourage them to take control in this way. The point is that legal liability should rest with those who have the capacity to make the necessary decisions. In the matter of catastrophic hazards this decision-making capacity is at company level.

If the mining companies were held responsible for safety failures they could be expected to play a more active role in ensuring that their mine managers attend to safety conscientiously. As things stand, operating companies can too easily adopt the view that it is the responsibility of the inspectorate to ensure that mines are operating safely.

There is a further consideration in the case of companies with complex corporate structures such as BHP. The legal liability of BHP Australia Coal in Brisbane will not necessarily impact on BHP head office in Melbourne. The attitude of head office is that prosecutions of the various BHP subsidiary companies are state matters and not of direct concern to BHP Ltd. However, in so far as BHP Ltd is ultimately responsible for safety, ways need to be found to hold it legally liable.

A final comment concerns the personal liability of top managers. The possibility of enforcing personal liability is an important element in the repertoire of an effective regulatory agency. General OHS legislation makes top managers liable to prosecution in certain circumstances, and the new Queensland coal mine legislation appears to follow suit.

This discussion has just skimmed the surface of the kinds of legal changes necessary to move legal liability further up the organisational hierarchy. It must be stressed again that all of this is quite academic unless the authorities are in a position to take resolute enforcement action. If they are not, the question of where legal liability lies is of little practical consequence.

SUMMARY

The move towards so-called self-regulation in the coal industry does not amount to a move in the direction of deregulation. The regime, as it is evolving, remains highly prescriptive. It is true that what is being prescribed is changing. The new guidelines for safety management plans prescribe the processes and parameters of these plans but not the precise content. In no significant sense is this a retreat from the field by government. However, if inspectorates do not have the resources or the political backing to compel companies to comply with these guidelines there is no guarantee that catastrophic risks will in fact be properly managed.

An important aspect of the move to self-regulation in Queensland is that it will place a duty of care firmly on the operating company. The statement by the senior BHP executive following the Moura disaster, that 'we accept accountability for the safety of our people and accept accountability for the disaster', was a statement without practical legal significance. BHP was at the time effectively immune from prosecution. Self-regulation in Queensland will mean that it is easier to hold companies legally accountable for disasters.

But, as is demonstrated by the situation in NSW where operators are already theoretically subject to duty of care legislation, the new legislation in Queensland will be inconsequential if the inspectorate is not in a position to *enforce* accountability—that is, to prosecute in the event that companies do not comply with their duty of care. The key issue in making the regulatory system more effective is not the style of regulation but the ability of inspectorates to enforce it. The current location of coal mine inspectorates in mining departments shackles them in this respect. Relocating them in larger agencies with the responsibility for enforcing health and safety legislation is one way in which these shackles might be removed.

12

CONCLUSION

Chapter 2 identified seven perspectives on disaster. Let us now reflect on how useful these have been in understanding the causes of the Moura tragedy. The perspectives overlap to varying degrees so we should not necessarily expect to be able to come to clear conclusions about which is most useful. Each in fact may sensitise us to different aspects of what went wrong. It is fairly clear, however, that the least applicable perspective is the first.

1 INEVITABILITY

Was the Moura disaster an inevitable outcome of work in a dangerous underground environment? That would seem to have been the view of the miner who made the first report. Recall his haunting words: 'We've lost her mate, we've bloody lost her.'

But the facts reveal that this disaster was far from inevitable. In the words of the inquiry report it was a result of 'management neglect and non-decision [making] which must never be repeated'. It was reasonably foreseeable and practically preventable. It could and should have been avoided.

2 THE PRIORITY OF PRODUCTION

The view that the Moura explosion stems from the priority given to production has much more to recommend it. Moura was a mine whose economic viability was in doubt and there was pressure to maximise production. As counsel for the miners put it in his final submission, 'BHP ran a dangerous race against time to extract all of the coal from 512 [panel] in accordance with the plan'. But, to be fair, it was not a race the company was consciously running. There was a whole culture of denial at Moura which served to nullify warnings of impending disaster. Production could continue at top speed only on the understanding that the mine was safe and this requirement generated an elaborate set of beliefs that provided this reassurance. This very same phenomenon was evident in the *Challenger* disaster where a belief evolved that O-ring malfunction was normal and acceptable. This belief served to disguise the fact that safety was being sacrificed so that launch schedules could be met.

There were numerous ways in which production was given a higher priority than safety at Moura. Most obviously, the routine practice of allowing men underground when a sealed panel was going through the explosive range involved a sacrifice of safety. More subtly, shift production reports were prepared at a higher level in the management structure than shift safety reports. Again, substantial financial incentives were paid for production but none for the conscientious management of catastrophic risk. This led to a situation where safety did not pay for relevant decision makers. Moura was not unusual in this respect: Bhopal and the *Herald of Free Enterprise* were two other disasters where it seemed that safety did not pay for relevant decision makers.

There is also a sense in which the very structure of government implicitly gives priority to production. I refer to the location of mine inspectorates in government departments whose primary function is the promotion of the mining industry. As we saw, this undermines the ability of inspectorates to impose safety standards on mines. This problem was identified in the Piper Alpha report and led to a recommendation to relocate the offshore inspectorate within the general occupational health and safety organisation in Britain, to ensure its independence from the industry it was supposed to regulate. This same structural problem is inherent in the regulatory regime for offshore petroleum production in Australia: safety is the responsibility of a government department whose primary concern is to facilitate production. Safety inspectorates in Australia's extractive industries need to be relocated to reduce the risk that safety will take second place to production.

3 Normal accident theory

As noted in Chapter 2, one of the better known disaster explanations is Perrow's theory of normal accidents. A normal or system accident, as Perrow defines it, involves the unanticipated interaction of multiple failures of various of the system components. Where a system is complex—that is, where its components can interact in unexpected ways—and where it is tightly coupled—that is, where multiple failures can occur so quickly that there is not sufficient time to analyse and react to what is happening—then system accidents, he says, are virtually inevitable.

It is clear that the Moura explosion cannot be analysed in these terms. The system was not complex in the above sense. On the contrary, there was nothing at all surprising or unexpected about the sequence of events; the process was quite foreseeable and had given rise to disaster on two occasions in Queensland in the recent past. Moreover, the system was not tightly coupled. Events unfolded slowly and resolute intervention at any point could have averted disaster and saved the mine. Perrow himself acknowledged that coal mine disasters could not be explained by his theory and the analysis here confirms his conclusion.

4 Misinformation

Turner's theory that 'disaster equals energy plus misinformation' turns out to have been a particularly fruitful approach to understanding the Moura disaster. According to Turner, disasters often occur because crucial information available at lower levels of the organisation is not assembled in a coherent way and is not passed on up the hierarchy to a point where it can be responsibly acted upon. At Moura, there were various reports of smells and of abnormally high CO readings. But there was no system for collating this information, or retrieving it for review, and so these isolated reports were rapidly lost.

The insights of high-reliability organisation theory are also relevant here. Organisations which are good at disaster avoidance mobilise as many pairs of eyes as possible to scrutinise critical information. This maximises the likelihood that the organisation will respond effectively to signs that things are going wrong.

This suggests the desirability of computerising information and transmitting it via e-mail or some other system to a range of people with relevant expertise at company headquarters and elsewhere. Computerisation would also mean that databases on warning signs could be maintained and patterns of warning signs easily identified.

Another feature of computerised systems is that they facilitate feedback to those who initially report information so that these reporters know how their reports have been received and have a chance to comment on whether the response is satisfactory. Had such feedback systems been in place at Moura it would have become clear to people making reports that their reports were being discounted and dismissed.

Turner suggests that the central question to ask is: 'What stops people from acquiring and using appropriate information?' The answers given above all concern the *system* by which the information was processed. But an unsatisfactory *system* is not the only reason why information failed to be used effectively. There were also cultural factors.

A hierarchy of knowledge operated at Moura that gave most weighting to information acquired by personal experience. Information acquired via oral communication was next in line while written communication was least effective. I have described this as a cultural phenomenon since it involved a belief system—an explicit belief that personal experience was the best guide and a belief that anything of importance in written reports would be transmitted orally. These beliefs systematically undermined the value of the written safety reports.

A second cultural phenomenon was the culture of denial—a set of beliefs that systematically discredited indicators that something might be wrong. They included the following beliefs: that the mine was safe unless it was proved otherwise; that abnormally high rates of CO were to be expected because of the novel mining method in use; that CO rates in any case had to show an exponential rise before there was cause for concern; that the mine was protected from spontaneous combustion by a six-month incubation period; and that no single indicator meant anything unless it was placed in context. This matrix of beliefs meant that the isolated and spasmodic indicators that something was wrong could be safely discounted.

To summarise: Turner's theory that disasters result from energy plus misinformation is highly relevant here. The Moura disaster occurred because vital information was rendered ineffective, both by an inadequate information processing system and by a culture that neutralised it.

5 Latent errors

Reason's theory of latent and active errors is also readily applicable. Recall his statement that major accidents occur when

> latent failures, arising mainly in the managerial and organisational spheres, combine adversely with local triggering events (weather, location etc) and with the active failures of individuals at the sharp end (errors and procedural violations).

This statement fits so well that it could have been written with Moura in mind. The latent failures at Moura were the many organisational problems identified in this study, especially the deficiencies in the system of communication. The local triggering event was the heating, and the active failures were the failures of individual managers to apply their minds conscientiously to the possibility that a heating might be occurring, particularly on the night of the explosion. Moura, in other words, is a classic illustration of the mechanisms that Reason identifies.

The particular significance of Reason's analysis is that it acknowledges that actions or inactions of individuals were a part of the chain of events that led to disaster, but concludes that focusing on these active failures of individuals is not as useful from the point of view of disaster prevention as focusing on the latent or organisational failings involved. The analysis, in a sense, provides a theoretical justification for the refusal of the Moura inquiry to dwell on the role of individuals in its account of what happened, as well as underpinning the organisational analysis undertaken in this book.

6 The failure of organisational memory

The theory that disasters occur because of a failure of organisational memory has particular significance in the Moura context. The mine that exploded in 1994 had not previously experienced an explosion, although two adjacent mines had. This mine therefore had no direct experience or memory to draw on. It might be argued that its management should have learnt from the disasters in adjacent mines, but that is another matter. The mine operator, BHP, did have experience of previous mine disasters, the most recent being the explosions at Moura No. 4 in 1986 and at Appin in NSW in 1979. It is BHP as an organisation whose memory was defective. As we have seen, this was because BHP had decentralised responsibility for disaster prevention to mine sites and had no central capacity or expertise with respect to disaster prevention. It is BHP as an organisation which could, in theory, develop a corporate memory and retain the lessons of the various disasters over which it has presided.

7 MANAGEMENT SYSTEM FAILURE

There is no doubt that the system for managing catastrophic risk at Moura was entirely inadequate. Management had not applied itself carefully and systematically to managing the problem of spontaneous combustion. It had, of course, given some thought to the matter. The 512 panel was designed so that extraction would be completed within three months, well inside the presumed incubation period of six months. Management had also installed a gas monitoring system and it was assumed that this would give warning of any spontaneous combustion that might be occurring. But no consideration had been given to just what would count as a warning and how the mine would respond to any such warning. In particular, no thought had been given to how the CO results would be interpreted and precisely what readings would trigger action. There was no action plan, that is, no set of procedures to be followed for dealing with a suspected spontaneous combustion. No serious thought had been given to how the mine would respond to the gas alarms generated by the CO monitoring system. Tragically, there were no procedures specified for the use of the sophisticated detection equipment purchased after the previous explosion. The result was that this equipment lay idle throughout the critical period. Finally, no thought had been given to just who would be responsible for doing what.

Management system failure is perhaps the most practical of all the perspectives considered in that it highlights management's role in planning for disaster prevention. No doubt for this reason it is the perspective emphasised by the Moura inquiry as well as other recent disaster inquiries. It led to the recommendation that mines develop specific plans for the management of a small number of specified catastrophic hazards—among them, spontaneous combustion—and it has led inspectorates in NSW and Queensland to draw up very detailed guidelines as to what mine managers should include in these plans.

Management system failure incorporates several of the other

perspectives considered above, notably Turner's misinformation, Reason's latent errors and Kletz's organisational memory failure. It does not supersede these, however, since it is these other perspectives that provide the insights into just how management systems fail.

ORGANISATIONAL FAILURES

The final sections of this chapter draw together the findings of this study in another way. The claim in Chapter 1 is that this is a study of how organisations fail. What follows is a summary of the failures we have identified; nearly all can be described as organisational failures.

- There was no adequate system for communication of decisions down the mine hierarchy.
- There was no adequate system of communicating information about warning signs up the hierarchy. In particular, people at BHP headquarters had no access to such information.
- There was no feedback mechanism to ensure that those reporting hazards or warnings were notified of what action, if any, was taken.
- There was a hierarchy of knowledge which led to a misplaced reliance on personal experience and on oral communication and a tendency to ignore written reports.
- The system for communicating production figures was more developed than the system for communicating safety information.
- There was a norm at the mine that unless warning signs were verified and confirmed they could be ignored.
- There was a series of beliefs at the mine, amounting to a culture of denial. One belief, in particular, 'normalised' the evidence—that is, it reinterpreted the indications of abnormality as lying within the bounds of normality.
- There were no specified actions that were mandatory when warning signs were detected.
- No one was identified as responsible for critical, safety-relevant decisions.
- The system of auditing by the company was thoroughly inadequate.
- The mine was using the wrong safety indicator—the lost time

injury frequency rate. This provided no indication of how well catastrophic risk was being managed.

- There were no incentives for mine managers to concern themselves with disaster prevention.
- BHP had decentralised responsibility for the control of catastrophic risk, along with all other aspects of safety, to the mine level. It had no corporate expertise in disaster prevention, despite its history of mine explosions.
- The law does not in practice hold the company or its senior management legally accountable for disaster prevention in the mine.
- Coal mines inspectorates are hamstrung in their enforcement efforts by their position in mining departments.

GETTING ORGANISATIONS TO ATTEND TO RISK

The whole problem can be summed up as one of organisational inattention—the attention of the organisation was simply not focused on the control of catastrophic risk. A number of principles emerge from this analysis about how to get organisations to attend to risk.

First, there is the matter of motivating managers at all levels. Senior managers must be made legally accountable for what goes wrong in their organisations. Furthermore, the structure of financial incentives must be changed so that managers are rewarded for attending conscientiously to the control of catastrophic risks.

Second, decision making in relation to the control of such risks must be structured so that the 'do nothing' option is removed. This involves the development of plans for the management of known major hazards. These plans must specify information-gathering processes, trigger events, what must be done when these trigger events occur, and who is responsible for carrying out these actions.

Third, there must be processes to guarantee that the plans are effective. They must be rigorously audited to ensure that they are working as intended. Auditors must find ways of pulling the triggers to see if what is intended to happen really does happen. Moreover, someone must audit the auditors to ensure that they are doing their job thoroughly. There is an important role here for the inspectorate. Inspectorates must also be willing to enforce the requirements of safety management plans and be prepared to prosecute when they find serious cases of non-compliance with plans that are supposed to be in operation.

The failures identified above and the principles of control to which they give rise are not specific to underground coal mining. They apply with perhaps some variations to the management of catastrophic risks generally. Similar failures occurring in organisations that do not have the potential for disaster nevertheless result in inefficiencies, irritations and suboptimal performance.

This book is a study of how organisations fail. It is also intended as a contribution to disaster prevention, for disasters are eminently preventable. Too often they amount to 'carbon copies' of earlier disasters (Braithwaite 1985:34–8). If only we had learnt and applied the lessons of those earlier disasters the most recent in the series would not have occurred. If BHP had attended to the lessons of earlier coal mine explosions, eleven men would not have lost their lives at Moura in 1994. The lessons of Moura must be learnt once and for all, so that this totally avoidable industrial carnage may cease.

Appendix 1

Company and Management Hierarchies

Note: This appendix reflects the situation in 1994. There has been some change since then.

Company hierarchy
BHP Company Ltd, owner of
BHP Minerals Pty Ltd, owner of
BHP Australia Coal Pty Ltd (BHPAC)

Management hierarchy within BHPAC and at Moura mine
BHPAC Group General Manager
BHPAC Regional General Manager, Southern Region
Manager of combined Moura open-cut and underground operations
Registered manager of Moura underground mine
Undermanager-in-charge
Shift undermanager
Deputy
Miner

APPENDIX 2

AUSTRALIAN MINE DISASTERS

This is a list of coal mine disasters in Australia involving ten or more fatalities.

	Year	*Number killed*
Bulli (NSW)	1887	81
Mt Kembla (NSW)	1902	96
Mt Mulligan (Qld)	1921	76
Bellbird (NSW)	1923	21
Box Flat (Qld)	1972	17
Kianga (Qld)	1975	13
Appin (NSW)	1979	14
Moura No. 4 (Qld)	1986	12
Moura No. 2 (Qld)	1994	11

REFERENCES

ACIL (1997). *Review of Mine Safety in NSW*. ACIL Economics & Policy, Brisbane.

Apostolou, D. (1997). 'BHP Coal and safety—the first priority'. *Proceedings of the Queensland Mining Industry Conference*. Department of Mines and Energy, Brisbane, pp. 195–203.

Appleton, B. (1994). Piper Alpha. In T. Kletz (ed.), *Lessons from Disaster: How Organisations have no Memory and Accidents recur*. Institute of Chemical Engineers, London, pp. 174–84.

Background Briefing, ABC Radio 25/4/95

Barrell, T. (1996a). *The Regulation of Health and Safety in the Australian Offshore Petroleum Industry*. Department of Primary Industries and Energy, Canberra.

Barrell, T. (1996b). *Review of the Management of Safety in the Offshore Operations of BHP Petroleum*. Department of Primary Industries and Energy, Canberra.

BASI (1993). *Report B/916/3032*. Bureau of Air Safety Investigation, Canberra.

BHPAC (1996). *Catastrophic Risk Management: What If . . .* BHPAC.

Braithwaite, J. (1985). *To Punish or Persuade: Enforcement of Coal Mine Safety*. State University of New York, Albany.

Brooks, A. (1994). *Guidebook to Australian Occupational Health and Safety Laws. 4th Edition*. CCH Australia, Sydney.

Carson, W. (1982). *The Other Price of Britain's Oil*. Martin Robertson, Oxford.

Chatterjee, P. (1980). *A Critical Review of Coal Mine Disasters in*

Queensland and New South Wales since 1920. Department of Mining and Metallurgy , University of Queensland.

Clarke, L. (1992). Context dependency and risk decision making. In J. Short & L. Clarke (eds), *Organisations, Uncertainties, and Risk.* Westview, Boulder.

Cullen, L. (1990). *The Public Inquiry into the Piper Alpha Disaster.* HMSO, London.

Cutler, T. & James, P. (1995). 'Moving beyond the side-car syndrome: a case of self-interest?' Paper given at a workshop on Understanding the Work Environment, Stockholm.

Cyert, R. & Marsh, J. (1992). *A Behavioural Theory of the Firm.* Blackwell Business, Cambridge, Mass.

Dalal, S., Fowlkes, E. & Hoadley, B. (1989). 'Risk analysis of the space shuttle: pre-Challenger prediction of failure'. *Journal of the American Statistical Association,* 84(408), 945–57.

Dawson, S., Willman, P., Clinton, A. & Bamford, M. (1988). *Safety at Work: the Limits of Self-regulation.* Cambridge University Press, Cambridge.

Department of Transport (1996). *Report on the incident to Boeing 737–400, G-OBMM near Daventry, on February 1995.* HMSO, London.

DPIE (1995). *Guidelines for Preparation and Submission of Safety Cases.* AGPS, Canberra.

Emmett, E. (1996). 'Prevention of major accidents in Australia'. *Journal of Occupational Health and Safety—ANZ,* 12(5), 581–9.

Etzioni, A. (1988). *The Moral Dimension: Towards a New Economics.* The Free Press, New York.

Festinger, L., Riecken, H. & Schachter, S. (1956). *When Prophecy Fails.* University of Minnesota, Minneapolis.

Glasbeek, H. & Tucker, E. (1993). 'Death by consensus: the Westray mine story'. *New Solutions* (Summer), 14–40.

Glen, H. (1993). 'Business responses to the regulation of health and safety in England'. *Law and Policy* 15(3), 219–33.

Gunningham, N. (1993). 'Environmental auditing: who audits the auditors?' *Environmental and Planning Law Journal* (August), 229–38.

Hopkins, A. (1995). *Making Safety Work: Getting Management Commitment to Occupational Health and Safety.* Allen & Unwin, Sydney.

Hudson, P., Reason, J., Wagenaar, W., Bentley, P., Primrose, M. & Visser, J. (1994). 'Tripod Delta: proactive approach to enhanced safety'. *Journal of Petroleum Technology* (January), 58–62.

Industry Commission (1995). *Work, Health and Safety, Vol 2.* AGPS, Canberra.

Johnson, H. T. & Kaplan, R. S. (1987). *Relevance Lost: The Rise and Fall of Management Accounting.* Harvard University Press, Boston.

Johnstone, G. (1991). Report of the State Coroner of Victoria into the Fire on the Esso Offshore Oil Platform, 'Tuna'.

Kletz, T. (1993). *Lessons from Disaster: How Organisations have no Memory and Accidents recur.* Institute of Chemical Engineers, London.

Kletz, T. (1994). *Learning from Accidents.* Butterworth-Heinemann, Oxford.

LaPorte, T. & Consolini, P. (1991). 'Working in practice but not in theory: theoretical challenges of "high-reliability organisations"'. *Journal of Public Administration Research and Theory* 1(1), 19–47.

Lepkowski, W. (1994). The restructuring of Union Carbide. In S. Jasanoff (ed.), *Learning from Disaster: Risk Management after Bhopal* University of Pennsylvania, Philadelphia, 22–43.

NOHSC (1994). *The Cost of Work-related Injury and Disease.* AGPS, Canberra.

NOHSC (1996). *Control of Major Hazard Facilities: National Standard.* AGPS, Canberra.

NSWDMR (1995 July). *Outburst Management Guidelines.* NSW Department of Mineral Resources, Sydney.

NSWDMR (1995 October). *Draft Spontaneous Combustion Management Code.* NSW Department of Mineral Resources, Sydney.

NSWDMR. (1996). *Explosion at Endeavour Colliery, 28 June 1995: Summary of Investigation.* NSW Department of Mineral Resources, Sydney.

Perrow, C. (1984). *Normal Accidents: Living With High-risk Technologies.* Basic, New York.

Perrow, C. (1994). 'The limits of safety: the enhancement of a theory of accidents'. *Journal of Contingencies and Crisis Management* 4(2), 212–20.

Purse, K. (1997). 'Workplace safety and microeconomic reform in Australia'. *International Journal of Employment Studies* 5(2).

Reason, J. (1990). *Human Error.* CUP, Cambridge.

Rees, J. (1994). *Hostages of Each Other: The Transformation of Nuclear Safety since Three Mile Island.* University of Chicago Press, Chicago.

Robinson, F. (1997). *A Study of Occupational Health and Safety Practice in the Construction of the Snowy Mountains Hydro-Electric Scheme: A Report to Worksafe*. National OHS Commission, Sydney.

Sagan, S. (1993). *The Limits of Safety*. Princeton University Press, Princeton.

Scholz, J. & Gray, W. (1990). 'A behavioural approach to compliance: OSHA enforcement's impact on workplace accidents'. *Journal of Risk and Uncertainty* 3, 283–305.

Seymour, C. (1996). Implementing the findings of the Moura enquiry. *Proceedings of the Futuresafe Conference*, Melbourne.

Simon, H. (1957). *Administrative Behaviour*. The Free Press, New York.

Standards Australia (1994). *Quality Assurance Explained*. Standards Australia, Sydney.

Stone, C. (1975). *Where the Law Ends: The Social Control of Corporate Behaviour*. Harper & Row, New York.

Turner, B. (1978). *Man-made Disasters*. Wykeham, London.

Vaughan, D. (1992). Regulating risk: implications of the Challenger accident. In J. Short & L. Clarke (eds), *Organisations, Uncertainties, and Risk*, Westview, Boulder, 235–53.

Vaughan, D. (1996). *The Challenger Launch Decision: Risky Technology, Culture and Deviance at NASA*. University of Chicago Press, London.

Voehl, F., Jackson, P. & Ashton, D. (1994). *ISO 9000: An Implementation Guide for Small to Mid-sized Businesses*. St Lucia Press, Florida.

Whiting, J. (1993). 'On safe behaviour'. *Australian Safety News* 64(7), 43–5.

Windridge, F. (1996). *Report on an Accident at Moura No. 2 Underground Mine on Sunday, 7 August 1994*. Brisbane: Warden's Court.

Windridge, F. (1996a). Coroner's report—Moura No. 2 fatal inquiry.

Woolfson, C. & Beck, M. (1995). 'The Piper Alpha disaster and the hidden deregulation of Britain's offshore oil industry'. Paper given at a workshop on Understanding the Work Environment, Stockholm.

Wright, C. (1994). 'A fallible safety system: institutionalised irrationality in the offshore oil and gas industry'. *The Sociological Review* 42(1), 79–103.

INDEX

www.ingramcontent.com/pod-product-compliance
Lightning Source LLC
LaVergne TN
LVHW021037011026
849160LV00009BA/677
* 9 7 8 1 8 6 5 0 8 7 0 2 3 *